Light

–

The Absolute Reference in the Universe

My motto:

When we study physical phenomena,
we always make
a mathematical model of them.

In such a model,
there are built-in physical laws
that are held together
by mathematical tools.

If the description of the physical phenomenon
is correct,
the mathematical model is
without contradictions
and
without paradoxes.

***

ex nihilo nihilum

Parmenides
500 BC

Jan Slowak

Light – The Absolute Reference in the Universe

Previous books

1. Bye-Bye Big Bang,
   Episod/Episode 1, 2, 3; swedish-english
4. Redshift factor, Absolute redshift,
   Galaxies red / blue distribution
5. Sawing of my article about the Big Bang
6. Big Bang –
   Questions to physicists and cosmologists
7. Einstein's special theory of relativity –
   mathematical and physical mistakes!
9. Back to Newton, swedish-english
10. Einstein's special theory of relativity =
    mathematical and ... swedish
11. SR(LT, LF, TD, LK) = NONSENS, swedish
12. Special Relativity is Nonsense, third edition

Publisher: BoD - Books on Demand, Stockholm, Sweden
Print: BoD - Books on Demand, Norderstedt, Germany
ISBN: 978-91-8007-066-9

For
Science

fourth edition

Content

Prologue

Can one question Einstein?

When discussing my previous book, I was asked if one can question Einstein. Consider the following: Aristotle was an adept for the geocentric worldview (Earth at the center of the solar system).

Quote from Wikipedia:
To explain the irregularities in the visible motions of the planets, Aristotle further developed Eudoxos' model with 27 concentric fictional spheres. Aristotle could not accept the spheres only as mathematical constructions. To him they must be real and then the transparent crystal spheres of the Pythagoreans came in handy. Significant spheres, however, complicated the model, which has already been refined by Kallippos into 33 spheres. The spheres were embedded in each other with the axes of the inner spheres mounted on the outer at different angles.
...

In any case, the epicycle theory could predict the orbits of the planets quite well.
Measurements / experiments verified this theory quite well.

Despite this, this theory was wrong because it was based on the wrong premise, that the earth is in the center of the solar system.

In the same way, it is said that experiments verify the special theory of relativity – SR!
I say that SR is based on incorrect grounds.
I point these out in my book:
Special Relativity is Nonsense.

I can say with my hand on my heart that one can question Einstein.
To help me, I would like to quote:
Cosmos – a brief history; Stephen Hawking;

"A physical theory is always provisional in the sense that it is only a hypothesis: it can never be proved. No matter how many times the experimental results correspond to a certain theory, one can still never be sure that the results next time contradict the theory. On the other hand, one can disprove a theory by finding only one that does not agree with the predictions of the theory "

1. Introduction

While I was researching the special theory of relativity, SR, various questions kept popping up:

- Relatively what is a reference system moving?
- Can an observer, trapped in an inertial reference system, determine whether one is moving or is at rest?
- Is moving relatively what?
- Is at rest relatively what?
- In what medium does the light and the other electromagnetic waves propagate?
- Is there ether or not?

We will analyze three different cases and try to answer some of these questions.

We will describe an inertial reference system in three different media:

Water
Air
Ether (interstellar space, vacuum)

Can we determine if we are in motion relative to these media or if we are in absolute rest in them?

We will use signals that are specific to each medium. These are the signals that two reference systems can use to communicate with each other.

The reader should remember that when we work with inertial reference systems and light signals, they should be considered as independent objects, independent phenomena.

On the other hand, one should keep in mind that in reality there are no inertial reference systems.
A reference system consists of matter and then the system is bound gravitationally to a celestial body. Then these systems move, for the most part, in circular, elliptical orbits. Inertial reference systems move in a line, at a constant speed.

Therefore, one should ask the question:
To which concrete physical phenomena can the special theory of relativity be applied?

2. Circles on the water

It is windless. The lake's water surface is completely still. There are no ripples on the water.
If we throw a small pebble, circular water waves will form. These move outwards from the point where the pebble reached the surface of the water. As these waves move toward the shore, their amplitude becomes smaller and smaller. Eventually the amplitude becomes so small that we will not be able to determine if the wave is still in motion.

No matter how many times we throw a pebble, this phenomenon will be repeated in the same way. Some waves will initially have a greater or lesser amplitude depending on the size of the pebble.
But they will move like concentric circles that get bigger and bigger.

But no matter what initial parameters we have in this experiment, these waves will move at the same speed, about 1 m/s.

In the next picture we see the lake's water surface and a boat. This is a simplified picture of reality. But when we make our mathematical model of reality, this picture will be even more simplified.

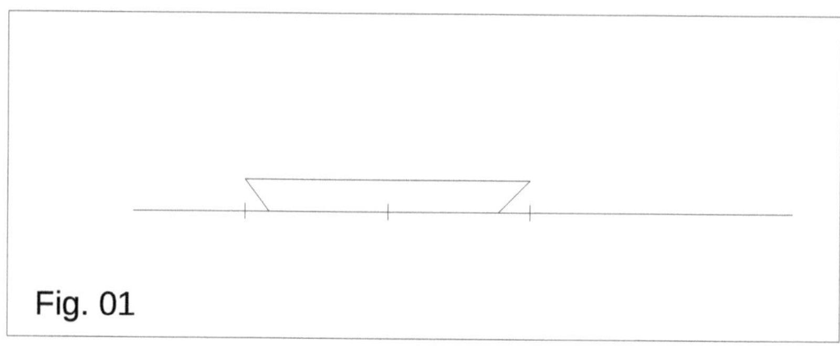

Fig. 01

For the next picture we take only the surface of the water which we represent with a line and the **front** of the boat F, **middle** M and the **rear** of the boat B. These three elements of the boat are shown as three points on the line.

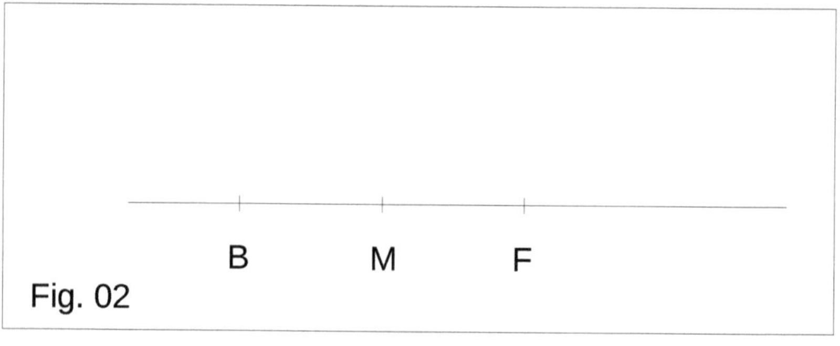

Fig. 02

We now imagine that a water wave is created at point M. It will move at the same speed both to the right, towards the front of the boat, and to the left towards the rear of the boat.

My question:
Can we only with the help of this knowledge determine whether an inertial reference system is at rest or in motion on the surface of the water?

My answer: Yes, we can.

Say that $BM = MF = d$.

Because the speed of the wave is the same in all directions, it will reach points B and F simultaneously. Always? No! Only if the boat (the reference system) is at rest relative to the water, the medium, in which the reference system is located and in which we study its motion.

The time that the water wave needs to reach the front of the boat and rear of the boat is

$$t_b = d / v_m; t_f = d / v_m$$

where d is the distance between the rear of the boat and the middle of the boat and between the front of the boat and the middle of the boat, v_m is the speed of the wave.

What happens if the boat moves to the right, forwards, at speed $v < v_m$?

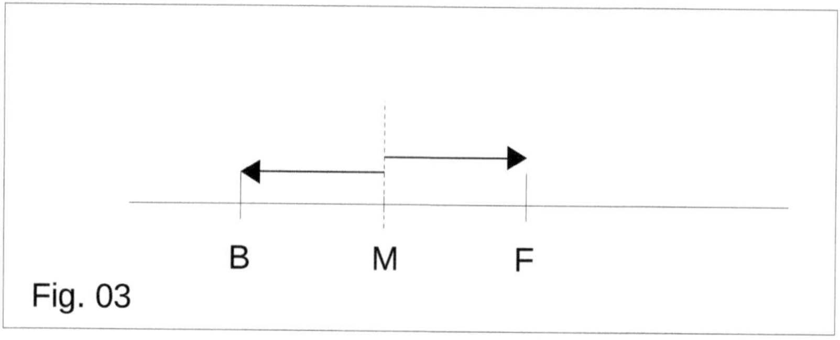

Fig. 03

While the wave will move towards the front of the boat and towards the rear of the boat, the boat will move forwards a short distance.

This means that the wave reaches first the rear of the boat and then the front of the boat. This is understood if the boat's speed is less than the wave's own speed which is always the same.

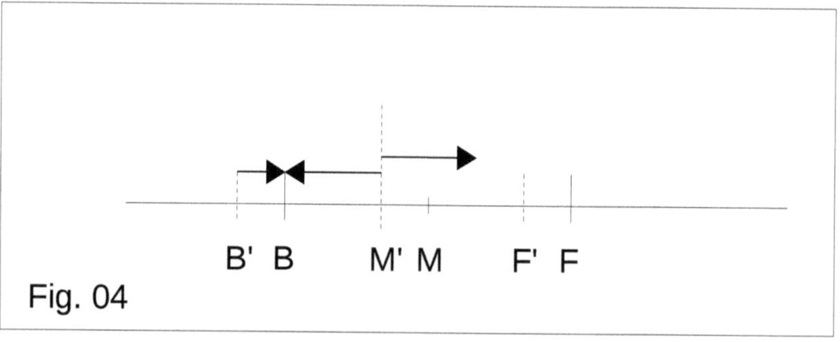

Fig. 04

We have marked with B', M', F' the boat's position when the wave occurred. $B'M' = BM = d$. We get the following relation:

$$d = B'B + BM' = tv + tv_m \rightarrow$$
$$d = tv + tv_m \rightarrow tv = d - tv_m \rightarrow$$
$$\boldsymbol{v = (d - tv_m)/t}$$

We read the time when the wave reaches the rear of the boat and can determine the boat's speed. If the read time is equal to

$$\boldsymbol{t = d/v_m} \rightarrow$$
$$v = (d - v_m\, d/v_m)/(d/v_m) \rightarrow$$
$$v = (d - d)/(d/v_m) \rightarrow$$
$$\boldsymbol{v = 0}$$

Now we look at what happens with the wave moving to the right, towards the front of the boat.

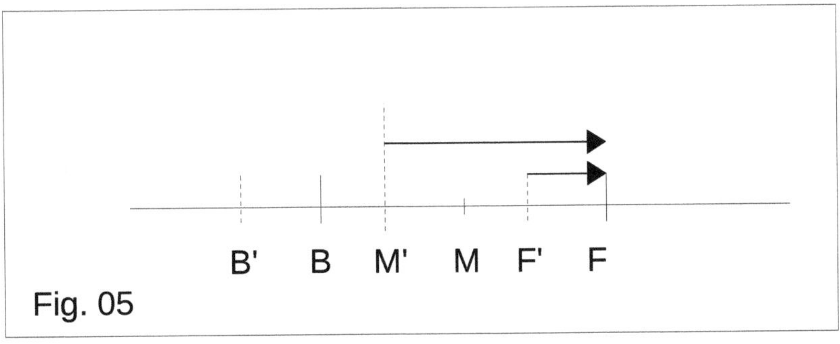

Fig. 05

In the picture above, we exclude the wave that moves to the left, towards the rear of the boat. We have already dealt with it.

Here we have the following: $M'F' = MF = d$. We get the following relation:

$$d = M'F - M'M = M'F - F'F = tv_m - tv \rightarrow$$
$$d = tv_m - tv \rightarrow tv = tv_m - d \rightarrow$$
$$\boldsymbol{v = (tv_m - d)/t}$$

We read the time when the wave reaches the front of the boat and can determine the speed of the boat. If the read time is equal to

$$\boldsymbol{t = d/v_m} \rightarrow$$
$$v = (tv_m - d)/t \rightarrow v = (v_m\, d/v_m - d)/(d/v_m) \rightarrow$$
$$v = (d - d)/(d/v_m) \rightarrow$$
$$\boldsymbol{v = 0}$$

We must remember that the three points in the water that represented the boat and the point where the wave occurred remain in the same place in the water reference system! The wave are moving! The boat moves or is still relative to the water!

3. The song of the birds

It is windless. The air temperature is comfortable. The weather is beautiful. We hear a bird singing. We see the bird in the distance. We can not determine if it is flying at a constant speed or if it is flapping its wings in the same place.

We want to do some experiments and see if we can determine if a bird is in motion relative to the air or if it is at absolute rest relative to the air. But bird has free will and can not be persuaded to participate in our experiments.

Then we use a drone.

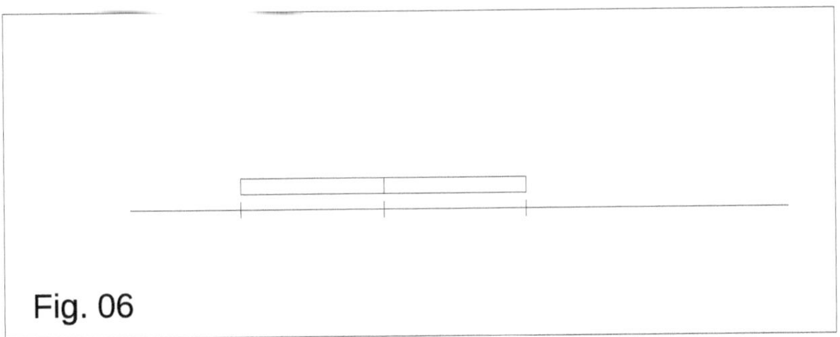

Fig. 06

Our drone has an elongated shape. In the middle is a small mechanism that can emit short beeps.

In the front and the rear there are receptors that can read these sound signals.

This is a simplified picture of reality but when we create a mathematical model to make calculations the image becomes even more abstract.
Then we show only three points on a line.

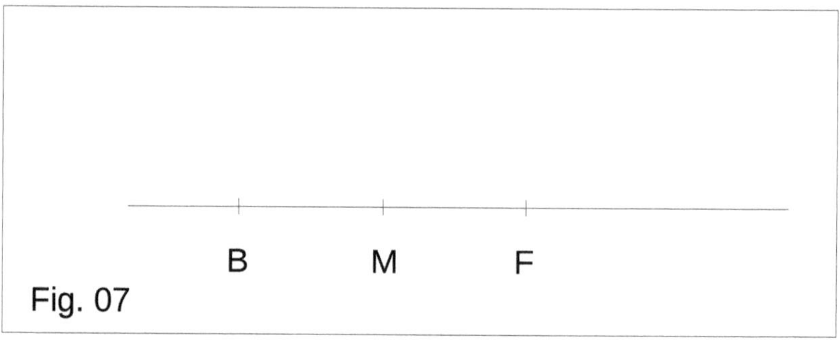

Fig. 07

The sound signal will propagate in all directions at the same speed. But we will consider only two points on the wavefront, one moving forwards (to the right of the image) and one moving backwards (to the left of the image). Under normal conditions, the speed of sound is about 343 m/s.

It is quite high speed to make measurements.
But we analyze this experiment from a theoretical point of view, from a mathematical point of view.
Because, in the moment we create the model of a physical phenomenon, it is only about mathematics!

The sound signal has different attributes but we only consider the speed of the sound. And we will try to determine if the object that emits the sound signal is in motion relative to the air or is in absolute rest relative to the air!

In the next picture we show the moment when the sound wave occurs in the middle and moves both forwards and backwards.

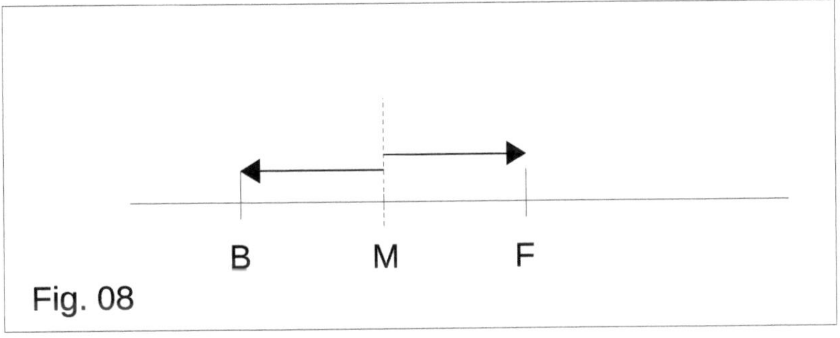

Fig. 08

My question:
Can we use this knowledge alone to determine whether an inertial reference system is at rest or in motion relative to air?

My answer: Yes, we can.

Say that $BM = MF = d$.

Because the speed of the wave is the same in all

directions, it will reach points B and F simultaneously. Always? No! Only if the drone (reference system) is at rest relative to the air, the medium, in which the reference system is located and in which we study its motion.

The time that the sound wave needs to reach the front and rear is

$$t_b = d / v_m; \; t_f = d / v_m$$

where d is the distance between the front and the middle of the drone and between the rear and the middle of the drone, v_m is the speed of the sound wave.

What happens if the drone moves to the right, forwards, at speed $v < v_m$?

While the sound wave will move towards the front and towards the rear, the drone will move forwards a short distance.

This means that the sound wave first reaches the rear and then the front. This is understood if the speed of the drone is less than the speed of the sound wave itself, which is always the same.

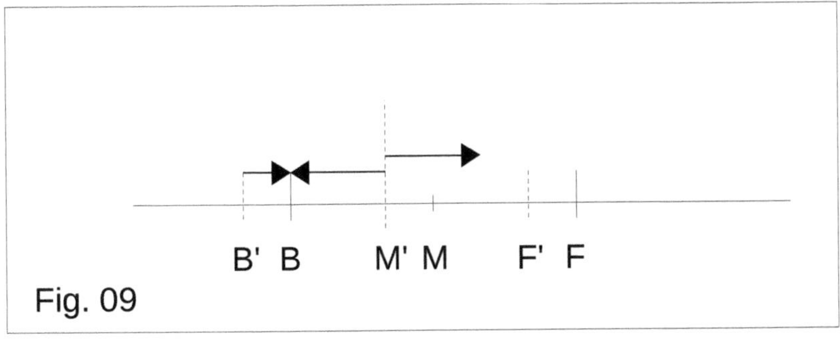

Fig. 09

We have marked with B', M', F' the position of the drone when the sound wave occurred. $B'M' = BM = d$. We get the following relationship:

$$d = B'B + BM' = tv + tv_m \rightarrow$$
$$d = tv + tv_m \rightarrow tv = d - tv_m \rightarrow$$
$$\boldsymbol{v = (d - tv_m)/t}$$

We read the time when the sound wave reaches the rear and can determine the drone's speed. If the read time is equal to

$$\boldsymbol{t = d/v_m} \rightarrow$$
$$v = (d - v_m\, d / v_m) / (d / v_m) \rightarrow$$
$$v = (d - d) / (d / v_m) \rightarrow$$
$$\boldsymbol{v = 0}$$

Now we look at what happens with the sound wave

moving to the right, towards the front. In the next picture, we exclude the sound wave that moves to the left, towards the rear. We have already dealt with it.

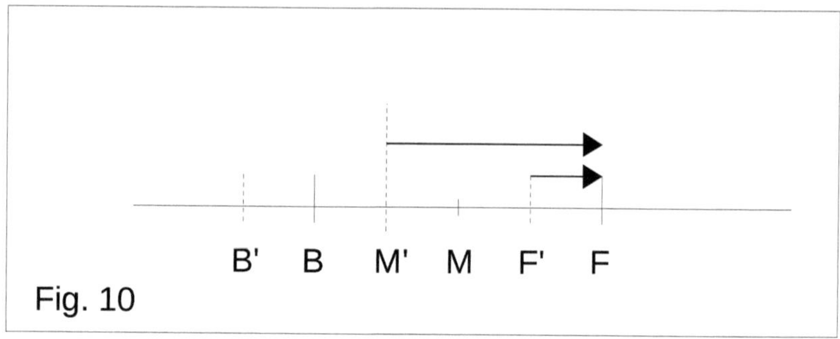

Fig. 10

Here we have the following: $M'F' = MF = d$. We get the following relation:

$$d = M'F - M'M = M'F - F'F = tv_m - tv \rightarrow$$
$$d = tv_m - tv \rightarrow tv = tv_m - d \rightarrow$$
$$\boldsymbol{v = (tv_m - d)/t}$$

We read the time when the sound wave reaches the front and can determine the drone's speed. If the read time is equal to

$$\boldsymbol{t = d/v_m} \rightarrow$$
$$v = (tv_m - d)/t \rightarrow v = (v_m\, d/v_m - d)/(d/v_m) \rightarrow$$
$$v = (d - d)/(d/v_m) \rightarrow$$
$$\boldsymbol{v = 0}$$

We must remember that the three points in the air that represent the drone and the point where the sound wave occurred remain in the same place in the air reference system! The sound wave is moving! The drone moves or is still relative to the air!

4. Alone among the stars

It is windless. It's night and the sky is cloudless. We know that the many stars we see are just a small part of all the stars in our galaxy, the Milky Way.

What would it be like to travel in a spaceship far away from our solar system? To calculate our arrival time to a certain star or a certain place in space, we must know our speed.
Speed relative to what?

Up among the stars, we can not use either water waves or sound waves to determine or calculate the speed of the spaceship! The only thing that can be used there in the interstellar vacuum are electromagnetic waves. Light is also an electromagnetic phenomenon. Its speed has been calculated by Maxwell to be about *300,000 km / s.*
Speed of light, but relative to what?

Then, in the middle of the 18th century, the medium in which light propagates was called ether. Light-bearing ether.

Then came the Michelson-Morley experiment.
Then Albert Einstein came up with the special theory

of relativity. And it was said that no ether was needed. That there is no ether.

Despite these claims, we know that light moves in the same way it did before this theory was written.
But we are far away among the stars and have nothing but light signals (think electromagnetic waves) to communicate with someone else or to determine if we are in motion or if we are in absolute rest.

Here we have our spaceship.

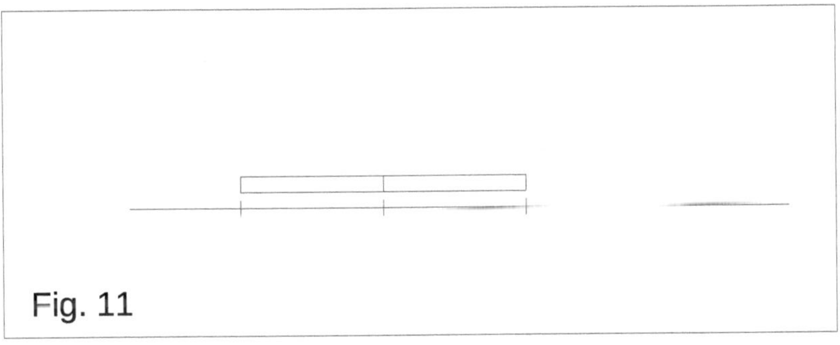

Fig. 11

Our spaceship has an elongated shape. In the middle is a small mechanism that can emit short light signals. In the front and the rear there are receptors that can read these light signals.

This is a simplified picture of reality but when we create a mathematical model to make calculations

the image becomes even more abstract.
Then we show only three points on a line.

We keep only the necessary elements to be able to do our experiments and make our calculations.

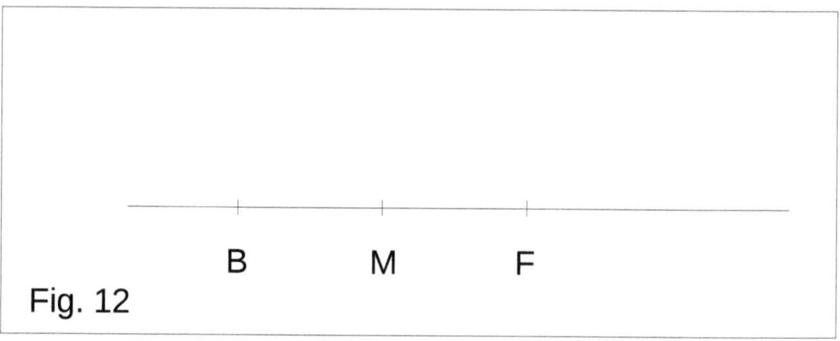

Fig. 12

The light signal will propagate in all directions at the same speed. But we will consider only two points on the wavefront, one moving forwards (to the right of the image) and one moving backwards (to the left of the image).
In the vacuum, the speed of light is
299.792,458 m / s.
We usually approximate it to *300,000 km / s*.

It is quite high speed to make measurements.
But we analyze this experiment from a theoretical point of view, from a mathematical point of view.
Because, in the moment we create the model of a physical phenomenon, it is only about mathematics!

The light signal has different attributes, but we only consider the speed of the light. And we will try to determine if the object that emits the light signal is in motion relative to the ether or is in absolute rest relative to the ether !

In the next picture we show the moment when the light wave occurs in the middle and moves both forwards and backwards.

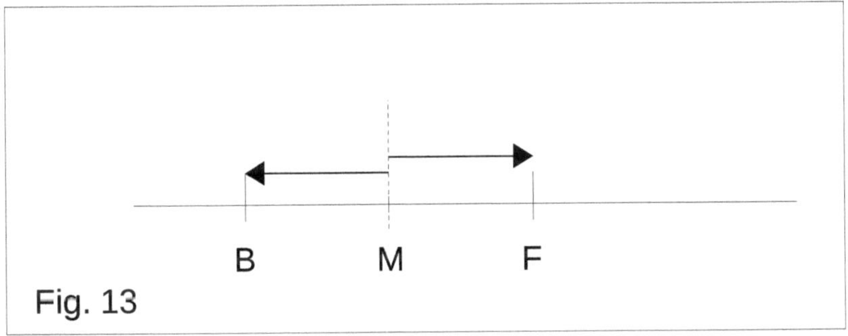

Fig. 13

My question:
Can we use this knowledge alone to determine whether an inertial reference system is at rest or in motion in the ether, in interstellar space?

My answer: Yes, we can.

Say that $BM = MF = d$.

Because the speed of the wave is the same in all

directions, it will reach points B and F simultaneously. Always? No! Only if the spaceship (the reference system) is at rest relative to the ether, the medium, in which the reference system is located and in which we study its motion.

The time the light wave needs to reach the front and rear is

$$t_b = d / v_m; \; t_f = d / v_m$$

where d is the distance between the front and the middle of the spaceship and between the rear and the middle of the spaceship, v_m is the speed of the light wave.

What happens if the spaceship moves to the right, forwards, at speed $v < v_m$?

While the light wave will move towards the front and towards the rear, the spacecraft will move forwards a small distance.

This means that the light wave reaches first the rear and then the front. This is understood if the speed of the spaceship is less than the speed of the light wave itself which is always the same.

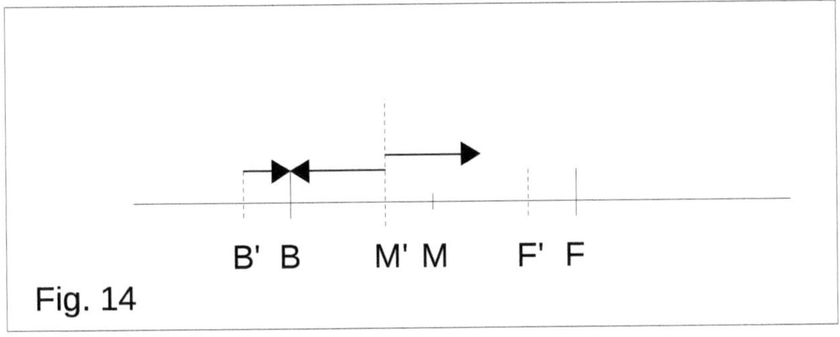

Fig. 14

We have marked with B', M', F' the position of the spaceship when the light wave occurred.
$B'M' = BM = d$. We get the following relationship:

$$d = B'B + BM' = tv + tv_m \rightarrow$$
$$d = tv + tv_m \rightarrow tv = d - tv_m \rightarrow$$
$$\boldsymbol{v = (d - tv_m)/t}$$

We read the time when the light signal reaches the rear of the spaceship and can calculate the spaceship's speed. If the read time is equal to

$$\boldsymbol{t = d/v_m} \rightarrow$$
$$v = (d - v_m\, d\,/\,v_m\,)\,/\,(d\,/\,v_m) \rightarrow$$
$$v = (d - d)\,/\,(d\,/\,v_m) \rightarrow$$
$$\boldsymbol{v = 0}$$

Now we look at what happens with the light wave

moving to the right, towards the front. In the next picture, we exclude the light wave that moves to the left, towards the rear. We have already dealt with it.

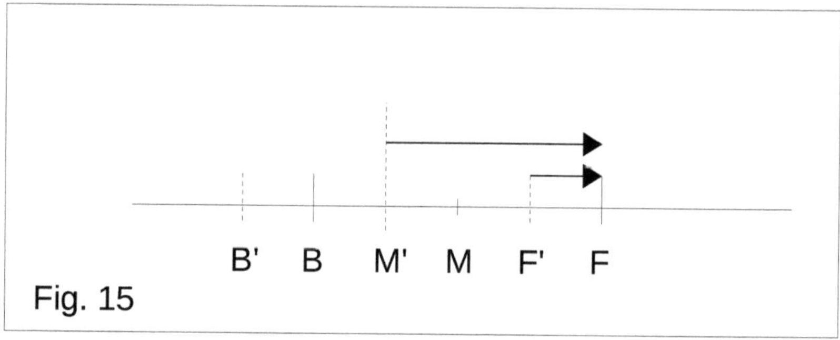

Fig. 15

Here we have the following: $M'F' = MF = d$.
We get the following relation:

$$d = M'F - M'M = M'F - F'F = tv_m - tv \rightarrow$$
$$d = tv_m - tv \rightarrow tv = tv_m - d \rightarrow$$
$$\boldsymbol{v = (tv_m - d)/t}$$

We read the time when the light wave reaches the front and can calculate the speed of the spaceship. If the read time is equal to

$$\boldsymbol{t = d/v_m} \rightarrow$$
$$v = (tv_m - d)/t \rightarrow v = (v_m d/v_m - d)/(d/v_m) \rightarrow$$
$$v = (d - d)/(d/v_m) \rightarrow$$
$$\boldsymbol{v = 0}$$

We must remember that the three points in the ether that represent the spaceship and the point where the light wave occurred remain in the same place in the ether's reference system! The light wave is moving! The spaceship moves or is still relative to the ether!

5. Three sides of the same coin

It is windless.

We have begun the previous three chapters with the same sentence. But that's not what we're talking about now.

I hope that the reader himself has come to the realization that what we have talked about in the previous three chapters is actually about the same thing.

We have analyzed whether an inertial reference system is in motion or at rest.

We have done this in:
- the water, where we used water waves
- the air, where we used sound waves
- the ether, where we used light waves.

We have used the same mathematical model.
We have used the same logic, the same calculations!

We have asked the same question:
Can we only with the help of this knowledge determine whether an inertial reference system is at rest or in

motion in the medium in which our reference system is located?

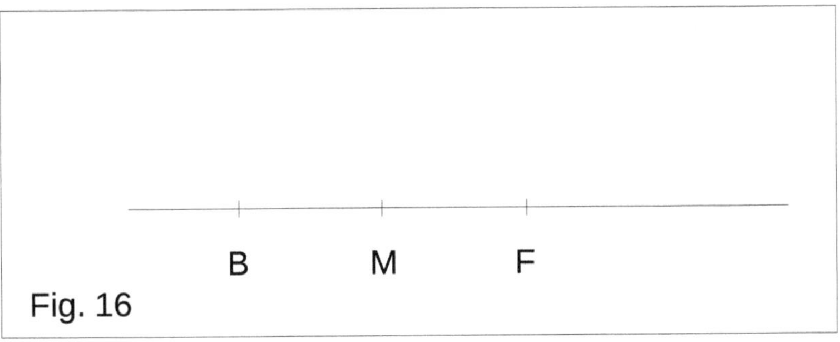

Fig. 16

My answer: Yes, we can.

The time that the wave needs to reach the front and rear is

$$t_b = d / v_m,\ t_f = d / v_m$$

where d is the distance between the front and the middle of the reference system and between the rear and the middle of the reference system, v_m is the velocity of the wave.

What happens if the reference system moves to the right, forwards, at speed $v < v_m$?

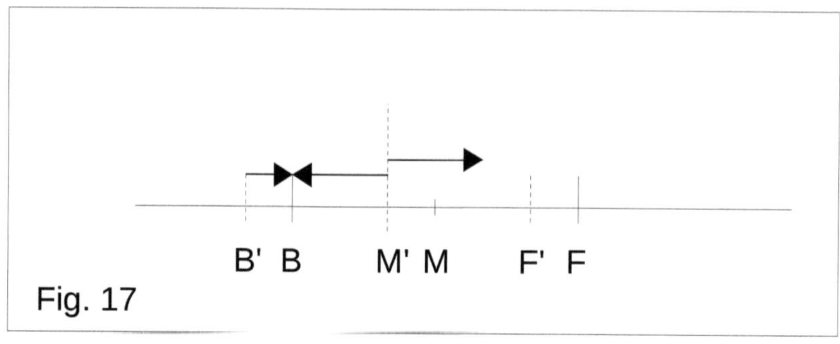

Fig. 17

Calculations:

$$d = B'B + BM' = tv + tv_m \rightarrow$$
$$d = tv + tv_m \rightarrow tv = d - tv_m \rightarrow$$
$$\boldsymbol{v = (d - tv_m)/t}$$

We read the time when the wave reaches the rear and can calculate the speed of the reference system. If the read time is equal to

$$\boldsymbol{t = d/v_m} \rightarrow$$
$$v = (d - v_m\, d / v_m) / (d / v_m) \rightarrow$$
$$v = (d - d) / (d / v_m) \rightarrow$$
$$\boldsymbol{v = 0}$$

Now we look at what happens with the wave moving to the right, towards the front.

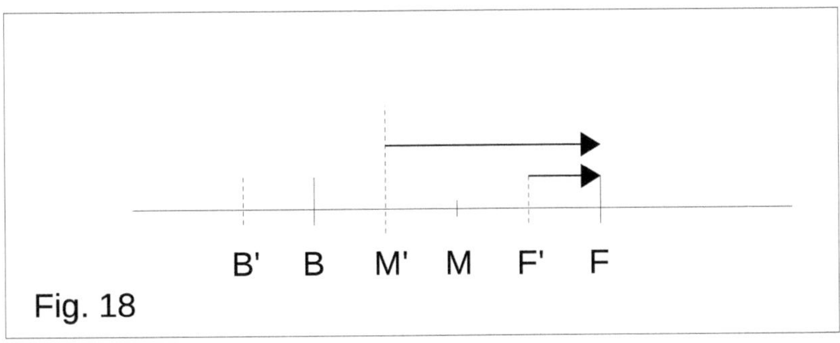

Fig. 18

Calculations:

$$d = M'F - M'M = M'F - F'F = tv_m - tv \rightarrow$$
$$d = tv_m - tv \rightarrow tv = tv_m - d \rightarrow$$
$$\boldsymbol{v = (tv_m - d)/t}$$

We read the time when the wave reaches the front and can calculate the speed of the reference system. If the read time is equal to

$$\boldsymbol{t = d/v_m} \rightarrow$$
$$v = (tv_m - d)/t \rightarrow v = (v_m\, d/v_m - d)/(d/v_m) \rightarrow$$
$$v = (d - d)/(d/v_m) \rightarrow$$
$$\boldsymbol{v = 0}$$

Above, we have presented three different variants of the same physical phenomenon.

The same mathematical model, the same logic, the same calculations, the same conclusions!

Furthermore, we will only talk about inertial reference systems and light signals.

6. Motion or absolute rest

We will touch on different concepts:
- coordinate system
- reference system
- inertial reference systems
- observer
- relative motion
- relative speed
- light signal
- absolute rest

Each time we talk about the relationship between these concepts, we will use a mathematical model.

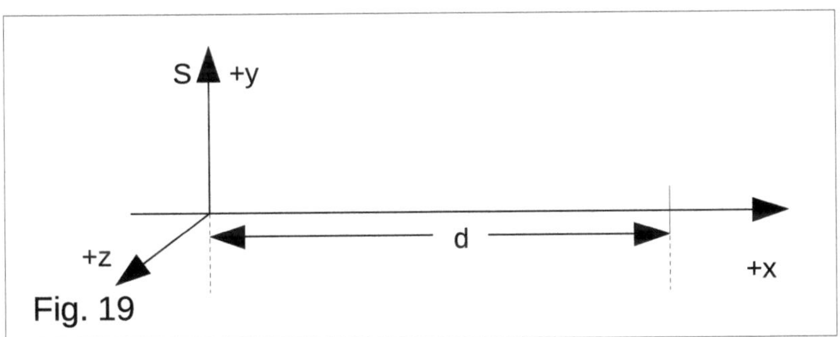

Fig. 19

In this model, we illustrate the following:
- a 3-dimensional Euclidean coordinate system (x, y, z), S
- a point on the x-axis which is at a distance *d* from the

S-origo.

For the most part, we will talk about two inertial reference systems and in order not to complicate the model, we will exclude unnecessary elements.

Fig. 20

Our models will not be drawn on a proportional scale due to the high speed of the light signal, about *300,000 km / s*.

We will illustrate how a reference system relates to a light signal using the figure above. This is about thought experiments.
The thought experiment begins when the clock in S shows the time zero.

At that moment, an event E, a light signal, occurs on the x-axis at a distance *d*.

We will determine the coordinates *x* and *t* for

this event as the reference system S read them at the moment when it becomes aware of the event, when the light signal reaches S.

We will always draw only the necessary elements in our figures so that you can easily perceive what is essential!
For example, we will not draw the distance between S and the event E.

We will not draw the y-axis and the z-axis because the event occurs on the x-axis and we will make calculations to find the distance between S-origo and point E where the event occurred.

We will do different thought experiments. For each experiment we will draw two or three figures:
- one for the initial position when the experiment begins
- one for the moment when the light signal reaches S'
- one for the moment when the light signal reaches S.

For each reference system, we will consider their velocity and its direction on the x-axis.
We'll talk about their absolute speed.

The literature says that one can only talk about the

relative speed of the system compared to another system. It is also claimed that it is not possible for an observer in a closed room, in an inertial reference system, to determine whether one is moving or whether one is at rest.

I claim it is possible and we have shown it in the previous chapters of this book and will illustrate this in the next section as well.

We consider the following inertial reference system:

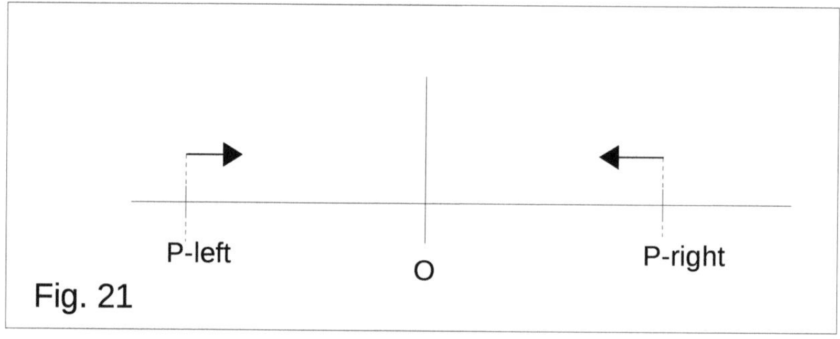

Fig. 21

From the points P-left and P-right, a light signal is sent to the O-observer.
When the experiment starts, the clock is reset to 0.
O registers the arrival time of the signals, *t_left* and *t_right*.
If these two times are equal, *t_left* = *t_right*, we can say that the reference system is at absolute rest relative to the x-axis.

The distance between P-left and point O is d, the distance between P-right and point O is d.

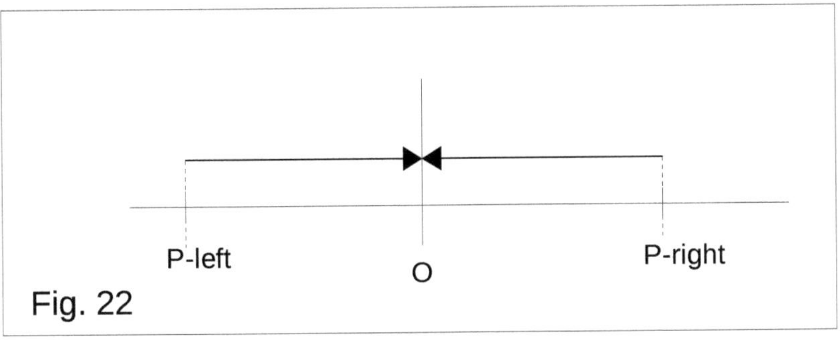

Fig. 22

$$t\_left = d\,/\,c, \quad t\_right = d\,/\,c$$

What happens if the reference system moves to the right at speed $v < c$?

The light signal from P-right will reach point O first.

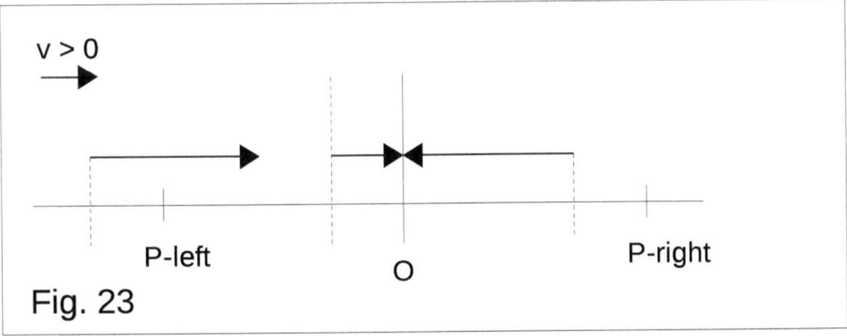

Fig. 23

$d = t\_right*c + t\_right*v \rightarrow d = t\_right(c + v) \rightarrow$

$\boldsymbol{t\_right = d/(c + v)}$

(* = multiplication sign)

After a while, the light signal from the P-left also reaches the point O.

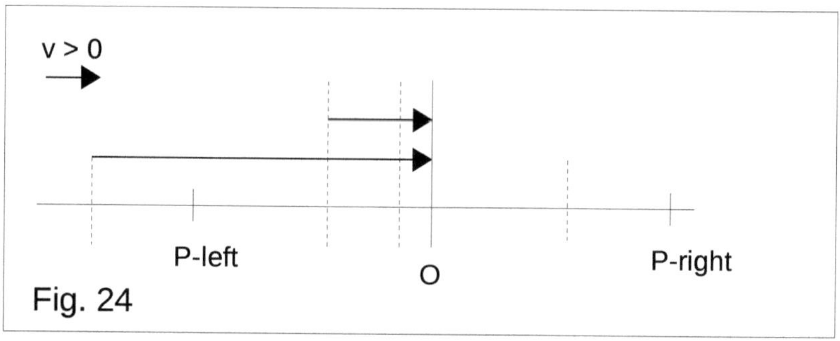

Fig. 24

$d = t\_left*c - t\_left*v \rightarrow d = t\_left(c - v) \rightarrow$

$\boldsymbol{t\_left = d/(c - v)}$

If the reference system move to the left, the two read times will be as in the next picture:

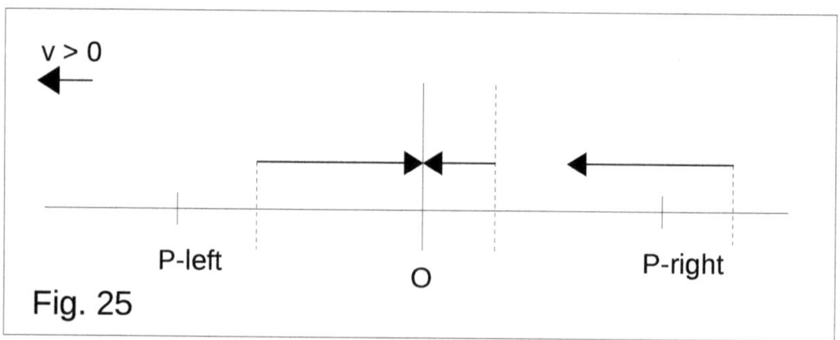

Fig. 25

The light signal from P-left will reach point O first.

$$d = t\_left*c + t\_left*v \rightarrow d = t\_left(c + v) \rightarrow$$
$$\mathbf{t\_left = d/(c + v)}$$

After a short while, the light signal from the P-right also reaches the point O.

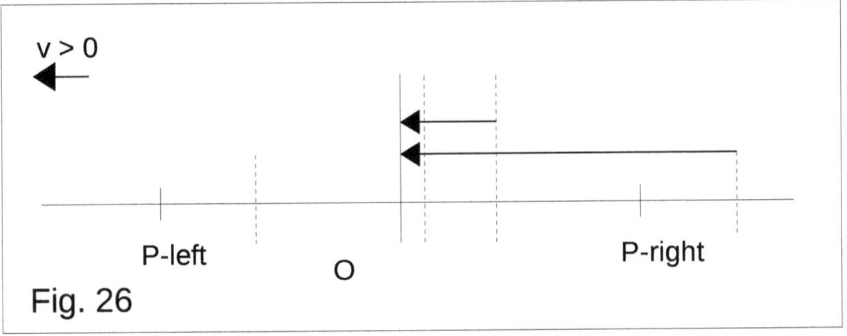

Fig. 26

$$d = t\_right*c - t\_right*v = t\_right(c - v) \rightarrow$$
$$\mathbf{t\_right = d/(c - v)}$$

We have illustrated 5 different cases and made calculations to get the two read times.

We can do this only if we know how the relative motion of the reference system on the x-axis is. And you do not know!

So we read the two times and draw the conclusion:
- if $t\_left = t\_right$ → the reference system is in **absolute rest** on the x-axis
- if $t\_left > t\_right$ → the reference system moves to the right
- if $t\_left < t\_right$ → the reference system moves to the left

In the next chapter, we deal with two inertial reference systems, their relationship to one and the same event. We define the event as a light signal that occurs on the x-axis.
We will make calculations of the positions of the reference systems on the x-axis, and the times they register.
We will consider these two reference systems so that they move at their absolute speed on x-axis.
They will always start from the same point, O.
Based on these considerations, we have identified the following positions that these two reference systems may have against each other.

				O		
TE1	S	S'				
TE2	S		S'			
TE3	S			S'		
TE4	S				S'	
TE5	S'	S				
TE6	S'		S			
TE7	S'			S		
TE8	S'				S	
TE9		S	S'			
TE10		S		S'		
TE11		S			S'	
TE12		S'	S			
TE13		S'		S		
TE14		S'			S	
TE15			S	S'		
TE16			S		S'	
TE17			S'	S		
TE18			S'		S	
TE19				S	S'	
TE20				S		S'
TE21				S'	S	
TE22				S'		S

Fig. 27

We see that some of these variants, experiments, are identical based on which end position, after a certain time $t > 0$, $t' > 0$, S and S' will have.
TE1 is identical to TE5, TE9, TE12
TE3 is identical to TE10
TE4 is identical to TE11
TE7 is identical to TE13
TE8 is identical to TE14
TE15 is identical to TE17
TE19 is identical to TE21

				O		
TE1	S	S'				
TE2	S		S'			
TE3	S			S'		
TE4	S				S'	
TE6	S'		S			
TE7	S'			S		
TE8	S'				S	
TE15			S	S'		
TE16			S		S'	
TE18			S'		S	
TE19				S	S'	
TE20				S		S'
TE22				S'		S

Fig. 28

We can eliminate the variants TE5, TE9, TE12, TE10, TE11, TE13, TE14, TE17, TE21.

We still have 13 variants and we will analyze these in more detail.
We are renumbering these remaining variants.

					O			
TE1	S	S'						
TE2	S		S'					
TE3	S				S'			
TE4	S						S'	
TE5	S'		S					
TE6	S'				S			
TE7	S'						S	
TE8				S	S'			
TE9				S			S'	
TE10				S'			S	
TE11						S	S'	
TE12						S		S'
TE13						S'		S

Fig. 29

In the next chapter, we analyze these variants. We regard them as thought experiments. Each experiment begins when the two reference systems are at point O. Then their clocks are reset. We consider their absolute velocity relative to point O in the vacuum, in the ether! When the clocks show $t = 0, t' = 0$, an event, a light signal, occurs at a point at a distance d from the point O.

7. Inertial reference systems – light signal from the right

Thought experiment R1

We consider two inertial reference systems, S and S'. At the beginning of the experiment, these two reference systems are located at the same point O on the x-axis. S moves to the **left** on the x-axis. S' is at **rest** relative to S, $\boldsymbol{v'} = \boldsymbol{v}$.
This means that S' also moves to the left at the same speed as S.

The point on the x-axis where the event occurs is denoted by the letter E, E′ (event).
Note that it is ONE POINT but we will consider it from two different reference systems, S and S′.
Therefore, we will denote the coordinates of the event with:

$E = (x, t)$ *for reference system S*
$E' = (x', t')$ *for reference system S′*

At the beginning of the experiment we have
$t = 0, t' = 0$.

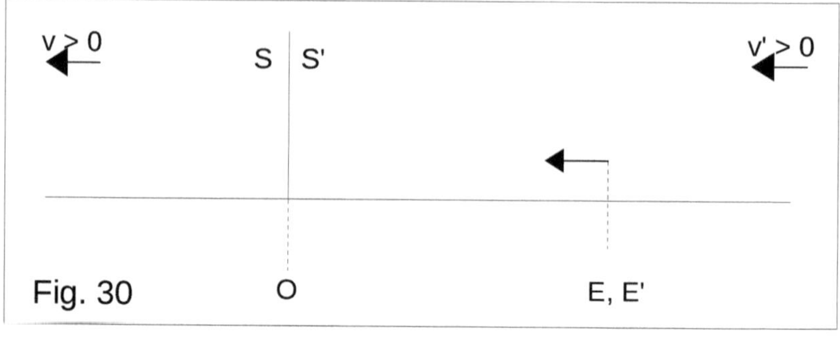

Fig. 30

At that moment, a light signal occurs on the x-axis. The light source is at a distance *d* from point O.

As the light signal moves towards S and S', S and S' move a short distance to the left. They move at the same speed.

The light signal will reach both reference systems simultaneously.

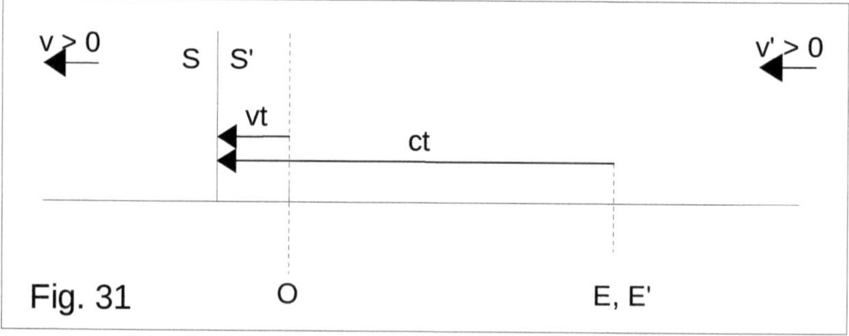

Fig. 31

We determine the coordinates of E.

$x = ct,\ d = ct - vt,\ d = t(c - v),\ t = d\,/\,(c - v) \rightarrow$
$\mathbf{E = (x, t) = (cd/(c - v),\ d/(c - v)).}$

It will be the same values for S'. In this case we have $x' = x,\ t' = t,\ v' = v.$

$x' = ct',\ d = ct' - v't',\ d = t'(c - v'),\ t' = d\,/\,(c - v') \rightarrow$
$\mathbf{E' = (x', t') = (cd/(c - v'),\ \ d/(c - v')).}$

Thought experiment R2

We consider two inertial reference systems, S and S'. At the beginning of the experiment, these two reference systems are located at the same point O on the x-axis. S moves to the left on the x-axis. S' moves to the left but at a lower speed than S, $v' < v$.

At the beginning of the experiment we have $t = 0, t' = 0$. At that moment a light signal occurs on the x-axis. The light source is at a distance d from point O.

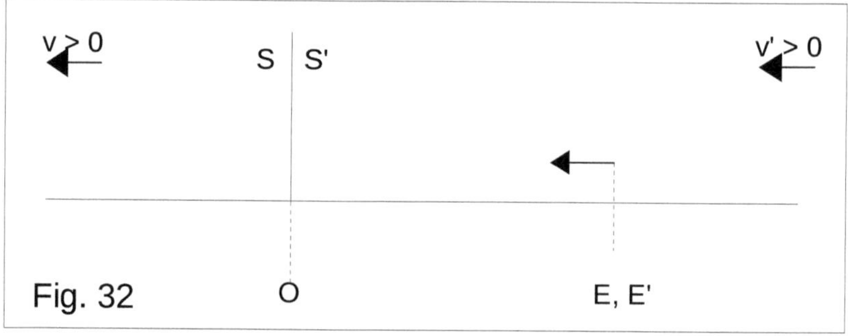

Fig. 32

As the light signal moves towards S and S', S and S' move to the left.

The light signal first reaches S'.

We determine the coordinates of E'.

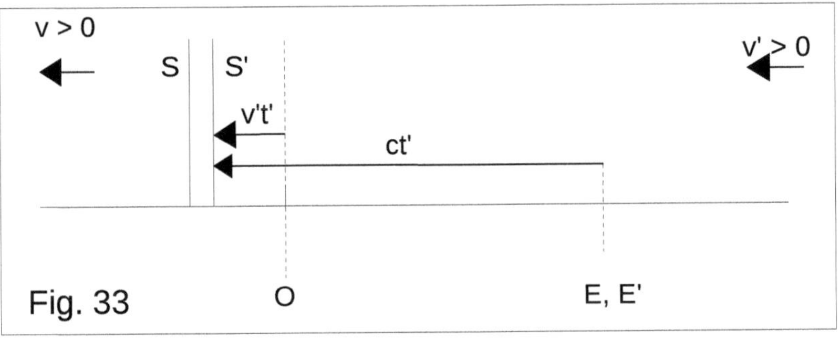

Fig. 33

$$x' = ct',\ d = ct' - v't',\ d = t'(c - v'),\ t' = d/(c - v') \rightarrow$$
$$\boldsymbol{E' = (x', t') = (cd/(c - v'),\ \ d/(c - v')).}$$

The light signal continues and will reach S. Meanwhile, S moves more to the left.

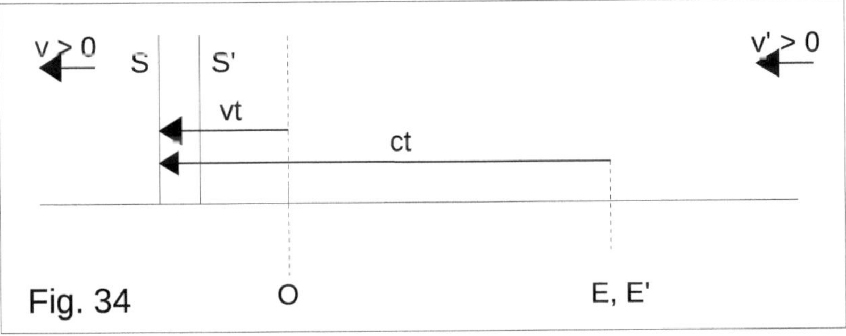

Fig. 34

We determine the coordinates of E.

$$x = ct,\ d = ct - vt,\ d = t(c - v),\ t = d/(c - v) \rightarrow$$
$$\boldsymbol{E = (x, t) = (cd/(c - v),\ \ d/(c - v)).}$$

Thought experiment R3

We consider two inertial reference systems, S and S'. At the beginning of the experiment, these two reference systems are located at the same point O on the x-axis.
S' is at **absolute rest** on the x-axis. S moves to **the left** relative to O.

At the beginning of the experiment we have $t = 0, t' = 0$. At that moment a light signal occurs on the x-axis. The light source is at a distance *d* from point O.

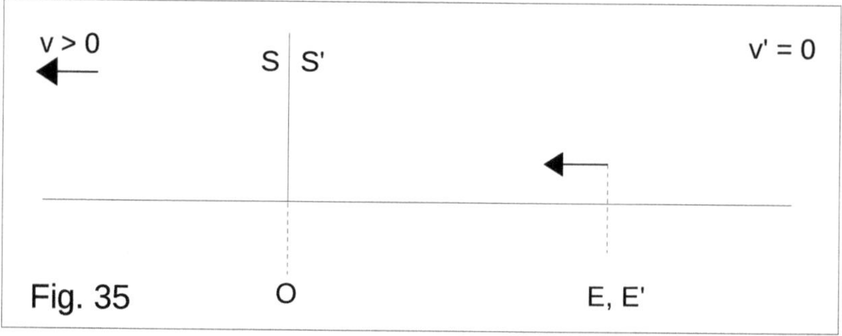

Fig. 35

While the light signal is moving towards S and S', S is moving a short distance to the left.

The light signal will first reach S'.

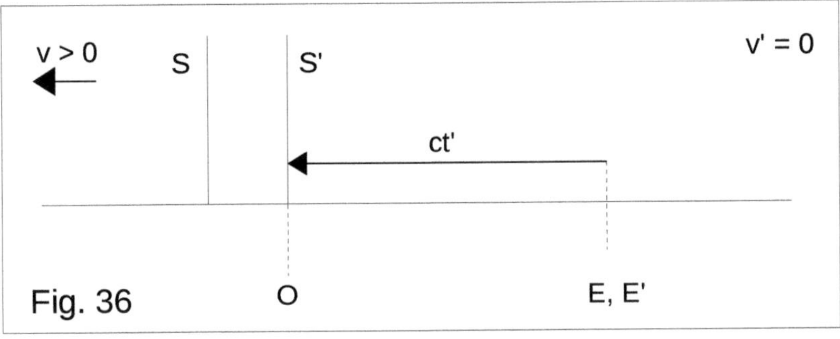

Fig. 36

We determine the coordinates of E'.

$$x' = ct',\ d = ct',\ t' = d/c \rightarrow$$
$$\boldsymbol{E' = (x', t') = (cd/c,\ \ d/c).}$$

The light signal continues and will reach S. We draw a separate line for the light signal that reaches S.

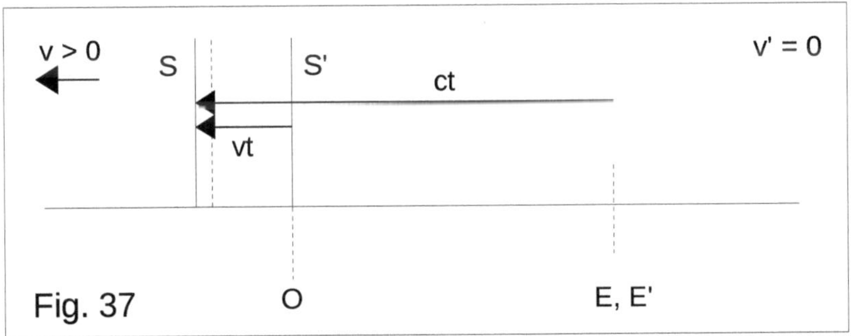

Fig. 37

We determine the coordinates of E.

$x = ct,\ d = ct - vt,\ d = t(c - v),\ t = d\,/\,(c - v) \rightarrow$
$\boldsymbol{E = (x, t) = (cd/(c - v),\ \ d/(c - v)).}$

Thought experiment R4

We consider **two** inertial reference systems, S and S'. At the beginning of the experiment, these two reference systems are located at the same point O on the x-axis. S moves to the **left** on the x-axis. S' moves to the **right** relative to O.

At the beginning of the experiment we have $t = 0, t' = 0$. At that moment a light signal occurs on the x-axis. The light source is at a distance d from point O.

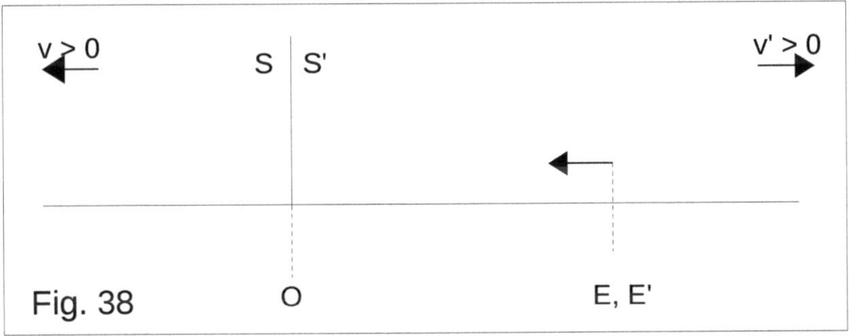

Fig. 38

As the light signal moves towards S and S', S moves to the left, S' to the right.

The light signal first reaches S'.

We determine the coordinates of E'.

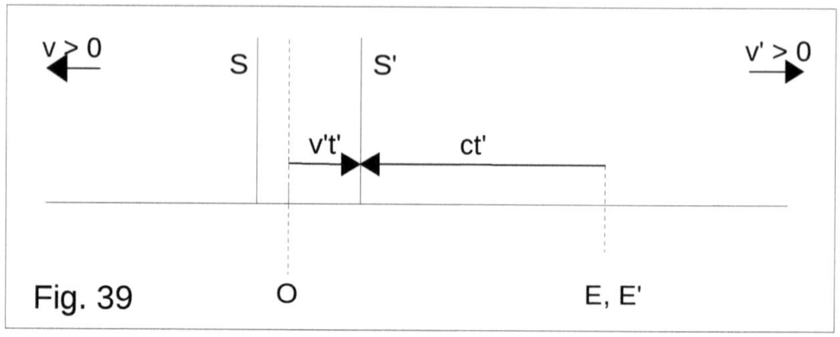

Fig. 39

$$x' = ct',\ d = ct' + v't',\ d = t'(c + v'),\ t' = d/(c + v') \rightarrow$$
$$\mathbf{E' = (x', t') = (cd/(c + v'),\ \ d/(c + v')).}$$

The light signal continues and will reach S. Meanwhile, S moves more to the left.

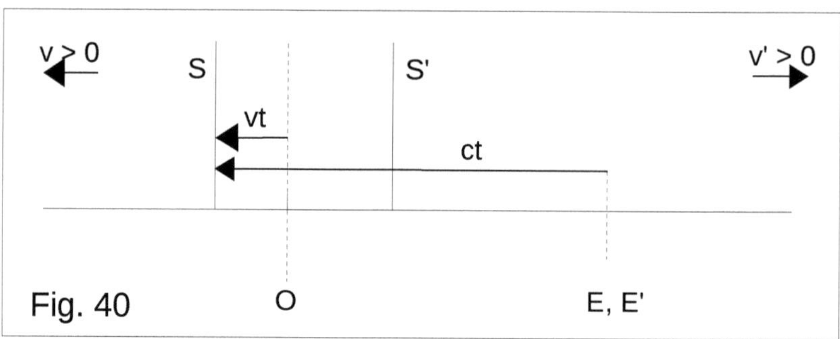

Fig. 40

We determine the coordinates of E.

$$x = ct,\ d = ct - vt,\ d = t(c - v),\ t = d/(c - v) \rightarrow$$
$$\mathbf{E = (x, t) = (cd/(c - v),\ \ d/(c - v)).}$$

Thought experiment R5

We consider **two** inertial reference systems, S and S'. At the beginning of the experiment, these two reference systems are located at the same point O on the x-axis. S moves to the **left** on the x-axis. S' moves to the **left** but at a higher speed than S, $v' > v$.

At the beginning of the experiment we have $t = 0, t' = 0$. At that moment a light signal occurs on the x-axis. The light source is at a distance d from point O.

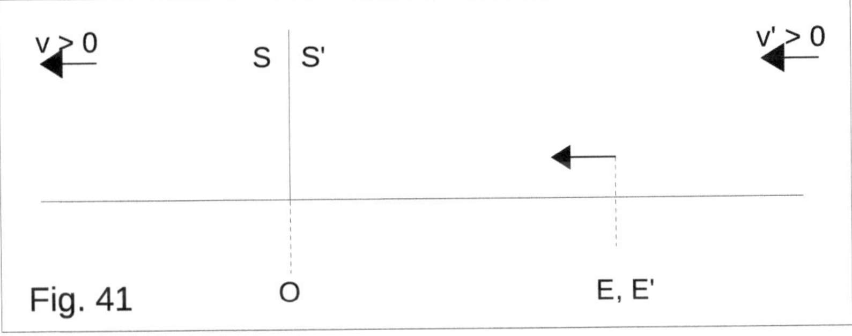

Fig. 41

As the light signal moves towards S and S', S and S' move to the left.

The light signal first reaches S.

We determine the coordinates of E.

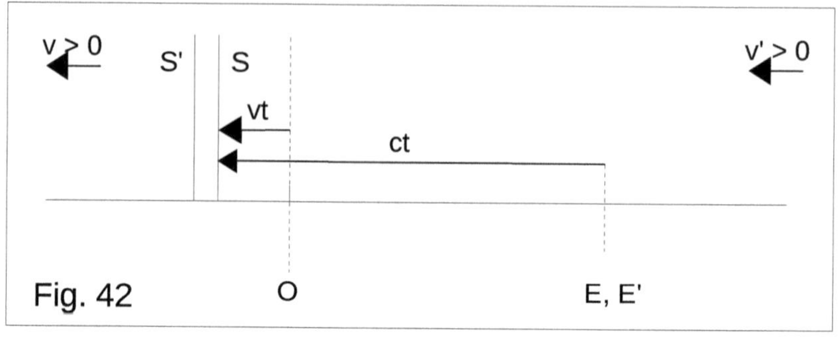

Fig. 42

$$x = ct,\ d = ct - vt,\ d = t(c - v),\ t = d\,/\,(c - v) \rightarrow$$
$$\boldsymbol{E = (x, t) = (cd/(c - v),\ \ d/(c - v)).}$$

The light signal continues and will reach S'. Meanwhile, S' moves more to the left.

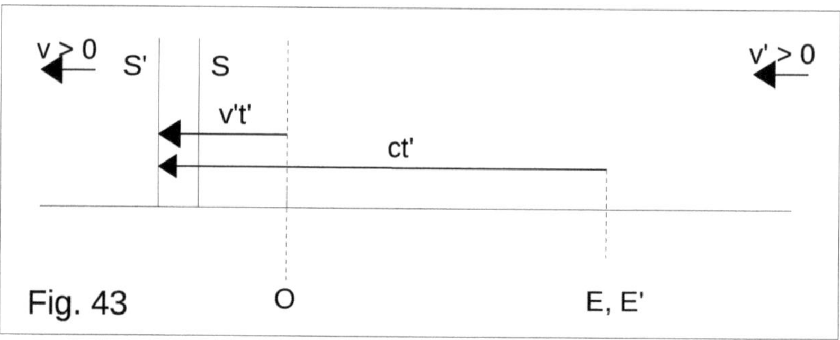

Fig. 43

We determine the coordinates of E'.

$$x' = ct',\ d = ct' - v't',\ d = t'(c - v'),\ t' = d\,/\,(c - v') \rightarrow$$
$$\boldsymbol{E' = (x', t') = (cd/(c - v'),\ \ d/(c - v')).}$$

Thought experiment R6

We consider **two** inertial reference systems, S and S'. At the beginning of the experiment, these two reference systems are located at the same point O on the x-axis.
S is at **absolute rest** on the x-axis. S' moves to the **left** relative to O.

At the beginning of the experiment we have
t = 0, t' = 0. At that moment a light signal occurs on the x-axis. The light source is at a distance *d* from point O.

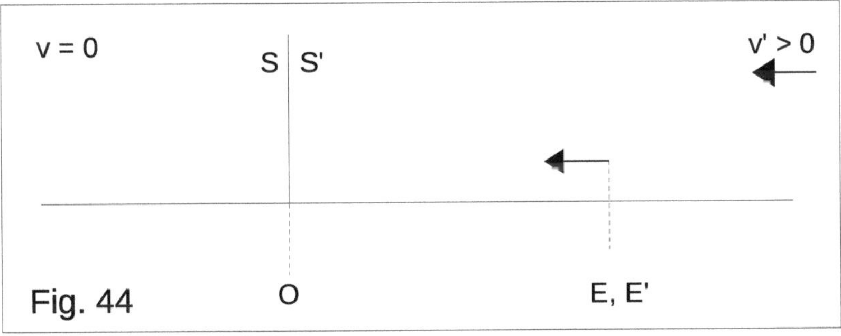

Fig. 44

While the light signal is moving towards S and S', S' is moving a short distance to the left.

The light signal will first reach S.

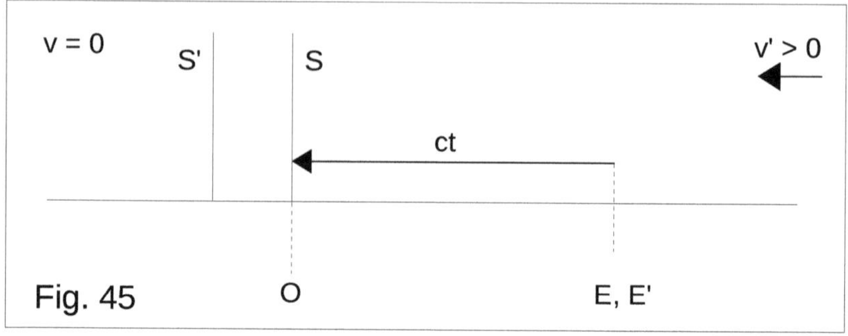

Fig. 45

We determine the coordinates of E.

$$x = ct,\ d = ct,\ t = d/c \quad \rightarrow$$
$$\boldsymbol{E = (x, t) = (cd/c,\ \ d/c).}$$

The light signal continues and will reach S'.

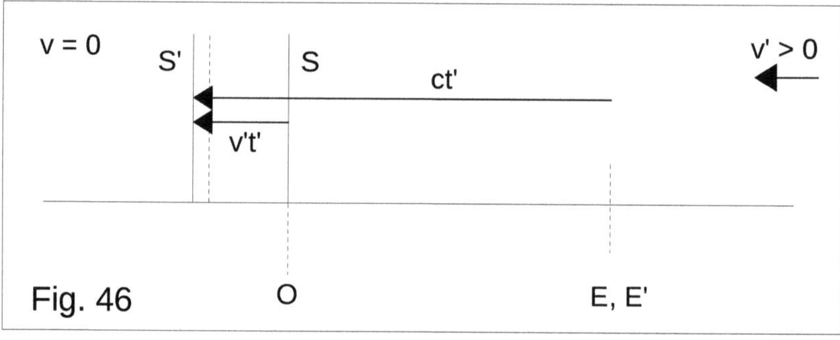

Fig. 46

We determine the coordinates of E '.

$$x' = ct',\ d = ct' - v't',\ d = t'(c - v'),\ t' = d/(c - v') \quad \rightarrow$$
$$\boldsymbol{E' = (x', t') = (cd/(c - v'),\ \ d/(c - v')).}$$

Thought experiment R7

We consider **two** inertial reference systems, S and S'. At the beginning of the experiment, these two reference systems are located at the same point O on the x-axis. S moves to the **right** on the x-axis. S' moves to the **left** relative to O.

At the beginning of the experiment we have $t = 0, t' = 0$. At that moment a light signal occurs on the x-axis. The light source is at a distance d from point O.

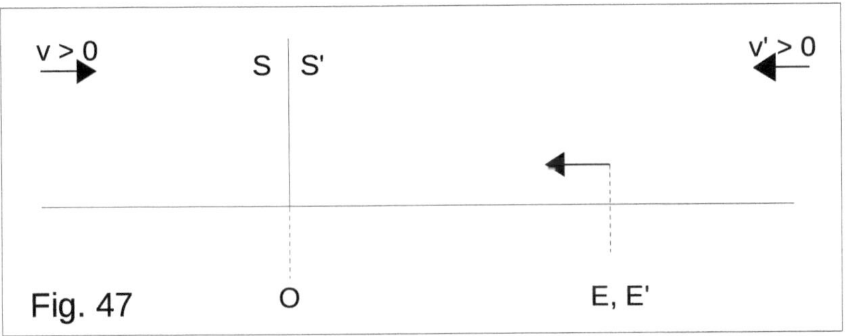

Fig. 47

As the light signal moves towards S and S', S moves to the right, S' to the left.

The light signal first reaches S.
We determine the coordinates of E.

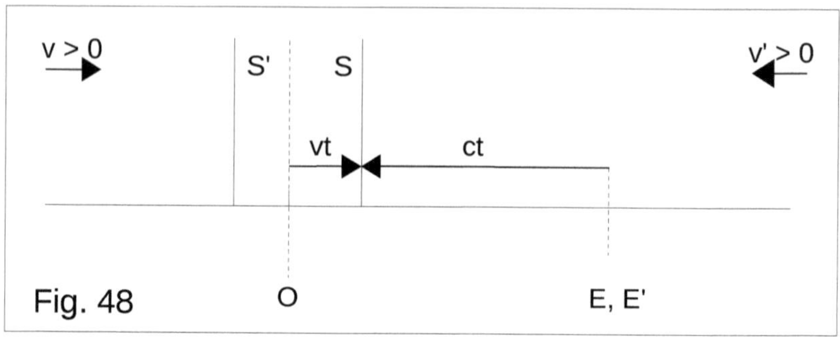

Fig. 48

$x = ct$, $d = ct + vt$, $d = t(c + v)$, $t = d/(c + v) \rightarrow$
$\boldsymbol{E = (x, t) = (cd/(c + v),\ d/(c + v))}$.

The light signal continues and will reach S'. Meanwhile, S' moves more to the left.

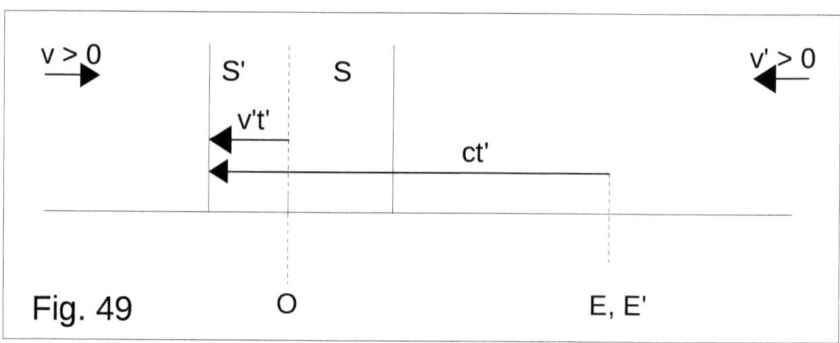

Fig. 49

We determine the coordinates of E'.

$x' = ct'$, $d = ct' - v't'$, $d = t'(c - v')$, $t' = d/(c - v') \rightarrow$
$\boldsymbol{E' = (x', t') = (cd/(c - v'),\ d/(c - v'))}$.

Thought experiment R8

We consider **two** inertial reference systems, S and S'. At the beginning of the experiment, these two reference systems are located at the same point O on the x-axis. These two reference systems are **at rest** relative to each other. And they are in **absolute rest** on the x-axis.

At the beginning of the experiment we have $t = 0$, $t' = 0$. At that moment a light signal occurs on the x-axis. The light source is at a distance *d* from point O.

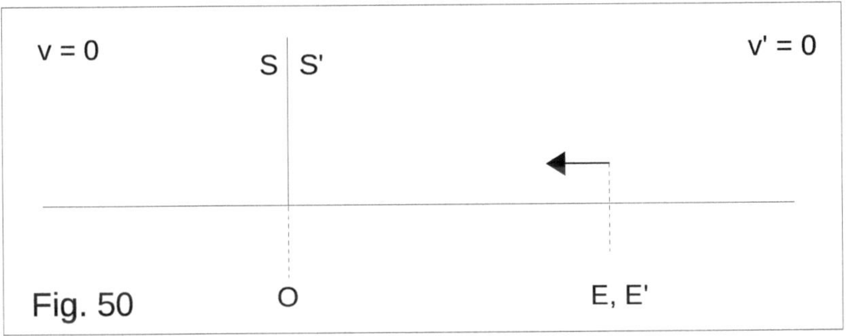

Fig. 50

When the light signal reaches the two reference systems, the times for the arrival of the light signal are registered.

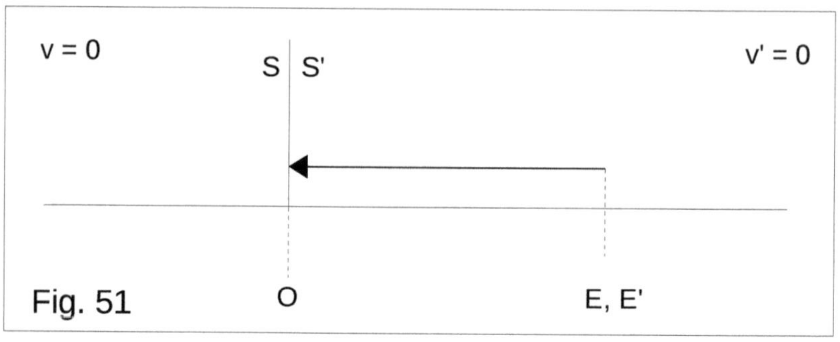

Fig. 51

We determine the coordinates of E.

$x = ct,\ d = ct,\ t = d / c \rightarrow$
$\boldsymbol{E = (x, t) = (cd/c,\ \ d/c).}$

In this case we have:

$x' = x$
$t' = t$
$v' = v = 0.$

We determine the coordinates of E'.

$x' = ct',\ d = ct',\ t' = d / c \rightarrow$
$\boldsymbol{E' = (x', t') = (cd/c,\ \ d/c).}$

Thought experiment R9

We consider **two** inertial reference systems, S and S'. At the beginning of the experiment, these two reference systems are located at the same point O on the x-axis. S is at **absolute rest** on the x-axis.
S' moves to **the right** relative to O.

At the beginning of the experiment we have $t = 0$, $t' = 0$. At that moment a light signal occurs on the x-axis. The light source is at a distance d from point O.

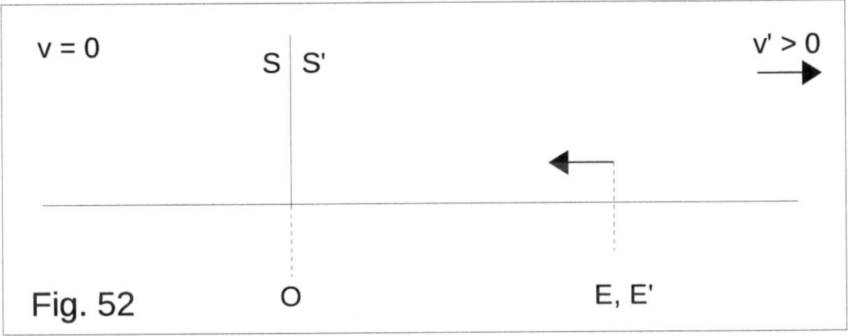

Fig. 52

While the light signal is moving towards S and S', S' is moving a short distance to the right.
The light signal will first reach S'.

We determine the coordinates of E' .

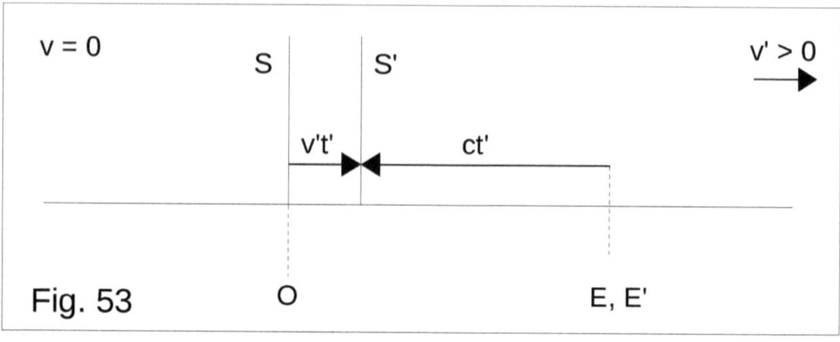

Fig. 53

$x' = ct'$, $d = ct' + v't'$, $d = t'(c + v')$, $t' = d/(c + v') \rightarrow$
$\boldsymbol{E' = (x', t') = (cd/(c + v'), d/(c + v'))}$.

The light signal continues and will reach S. Meanwhile, S' moves a little further to the right. We draw a separate line for the light signal that reaches S.

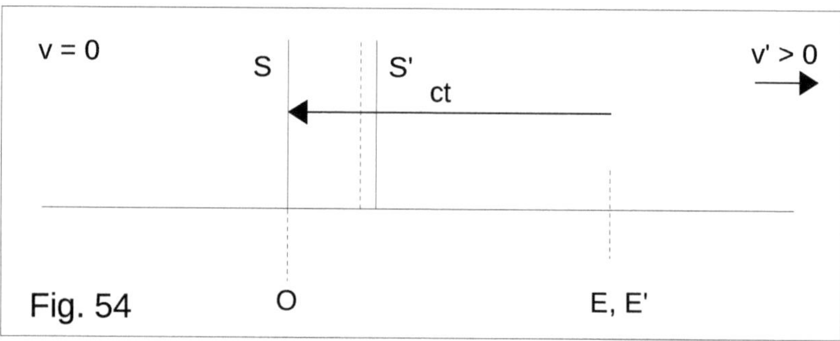

Fig. 54

We determine the coordinates of E.

$x = ct$, $d = ct$, $t = d/c \rightarrow$
$\boldsymbol{E = (x, t) = (cd/c, d/c)}$.

Thought experiment R10

We consider **two** inertial reference systems, S and S'. At the beginning of the experiment, these two reference systems are located at the same point O on the x-axis. S' is at **absolute rest** on the x-axis.
S moves to **the right** relative to O.

At the beginning of the experiment we have $t = 0$, $t' = 0$. At that moment a light signal occurs on the x-axis. The light source is at a distance d from point O.

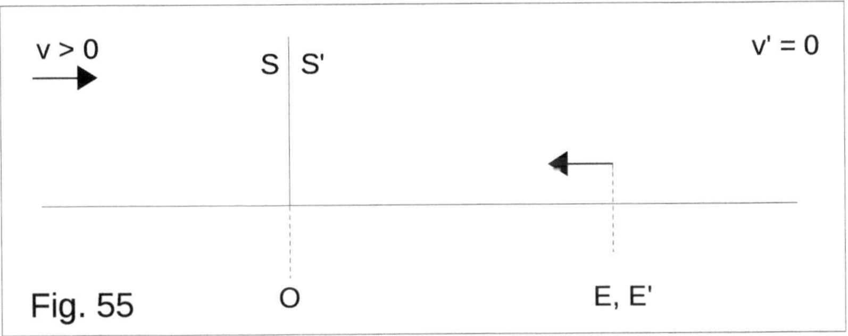

Fig. 55

While the light signal is moving towards S and S', S is moving a short distance to the right.
The light signal will first reach S.

We determine the coordinates of E.

$$x = ct,\ d = ct + vt,\ d = t(c + v),\ t = d\,/\,(c + v) \rightarrow$$

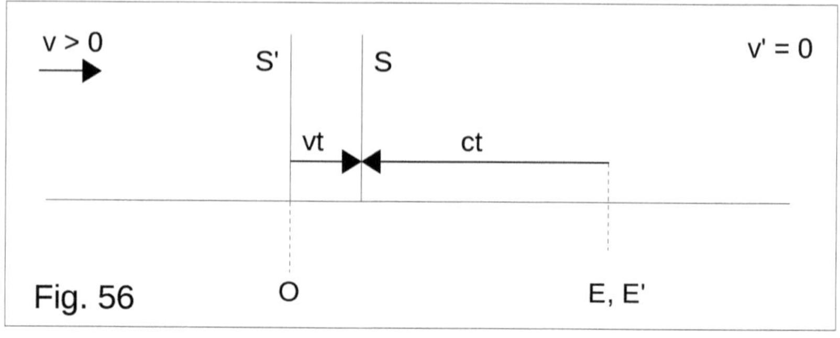

Fig. 56

$$\boldsymbol{E = (x, t) = (cd/(c + v), d/(c + v)).}$$

The light signal continues and will reach S'. Meanwhile, S moves a little further to the right. We draw a separate line for the light signal that reaches S'.

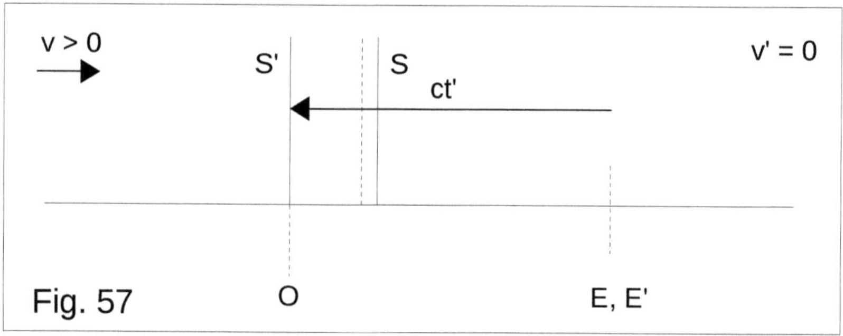

Fig. 57

We determine the coordinates of E'.

$$x' = ct',\ d = ct',\ t' = d/c \quad \rightarrow$$
$$\boldsymbol{E' = (x', t') = (cd/c,\ \ d/c).}$$

Thought experiment R11

We consider two inertial reference systems, S and S'. At the beginning of the experiment, these two reference systems are located at the same point O on the x-axis. S moves to **the right** on the x-axis. S' is at **rest** relative to S, $v' = v$.
This means that S' also moves to the right at the same speed as S.

At the beginning of the experiment we have $t = 0, t' = 0$. At that moment a light signal occurs on the x-axis. The light source is at a distance d from point O.

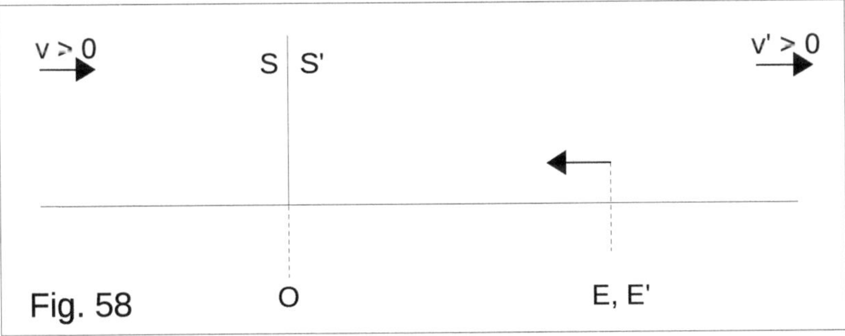

Fig. 58

As the light signal moves towards S and S', S and S' move a short distance to the right. They move at the same speed.

The light signal will reach both reference systems simultaneously.

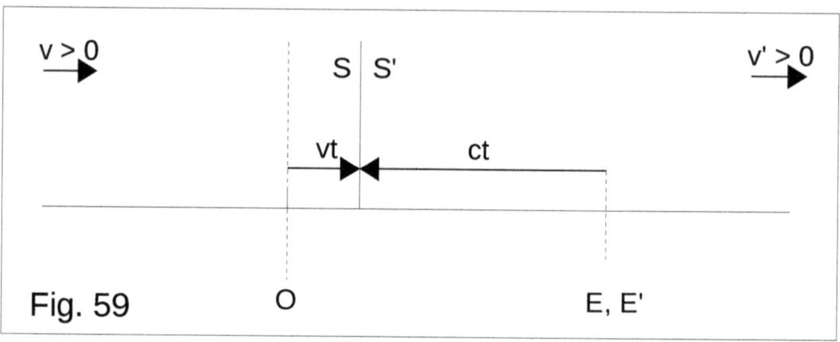

Fig. 59

We determine the coordinates of E.

$x = ct$, $d = ct + vt$, $d = t(c + v)$, $t = d/(c + v) \rightarrow$

$$\boldsymbol{E = (x, t) = (cd/(c + v),\ d/(c + v))}.$$

It will be the same values for E'. In this case we have $x' = x$, $t' = t$, $v' = v$.

$x' = ct'$, $d = ct' + v't'$, $d = t'(c + v')$, $t' = d/(c + v') \rightarrow$

$$\boldsymbol{E' = (x', t') = (cd/(c + v'),\ d/(c + v'))}.$$

Thought experiment R12

We consider **two** inertial reference systems, S and S'. At the beginning of the experiment, these two reference systems are located at the same point O on the x-axis. S moves to **the right** on the x-axis.
S' moves to **the right** but at a higher speed than S, $v' > v$.

At the beginning of the experiment we have $t = 0, t' = 0$. At that moment a light signal occurs on the x-axis. The light source is at a distance *d* from point O.

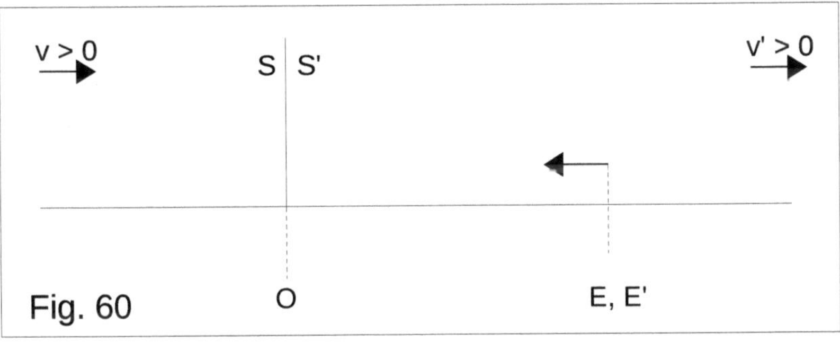

Fig. 60

As the light signal moves towards S and S', S and S' move a short distance to the right. S' moves faster than S.

The light signal will first reach S'.

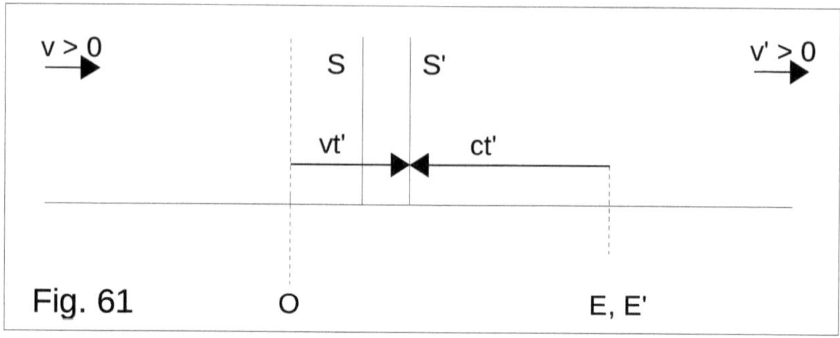

Fig. 61

We determine the coordinates of E'.

$x' = ct'$, $d = ct' + v't'$, $d = t'(c + v')$, $t' = d\,/\,(c + v') \rightarrow$
$\boldsymbol{E' = (x', t') = (cd/(c + v'),\ \ d/(c + v'))}$**.**

The light signal continues and will reach S. Meanwhile, S and S' move a little further to the right.

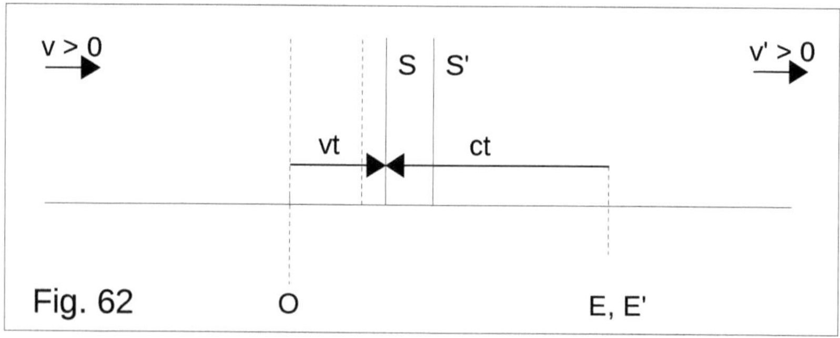

Fig. 62

We determine the coordinates of E.

$x = ct$, $d = ct + vt$, $d = t(c + v)$, $t = d\,/\,(c + v) \rightarrow$
$\boldsymbol{E = (x, t) = (cd/(c + v),\ \ d/(c + v))}$**.**

Thought experiment R13

We consider two inertial reference systems, S and S'. At the beginning of the experiment, these two reference systems are located at the same point O on the x-axis. S moves to the right on the x-axis. S' moves to the right but with less speed than S, $v' < v$.

At the beginning of the experiment we have $t = 0, t' = 0$. At that moment a light signal occurs on the x-axis. The light source is at a distance d from point O.

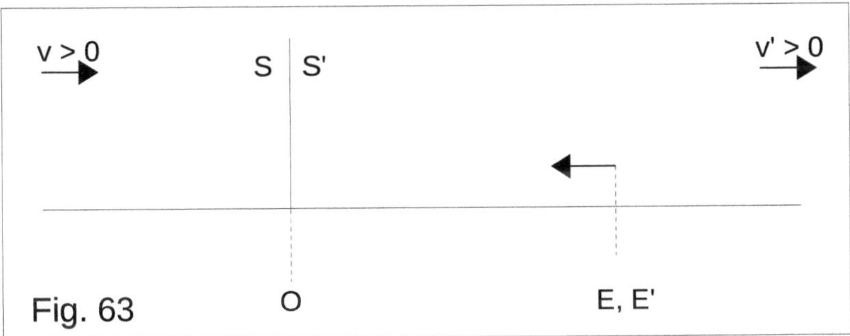

Fig. 63

As the light signal moves towards S and S', S and S' move a short distance to the right.

S moves faster than S'.
The light signal will first reach S.

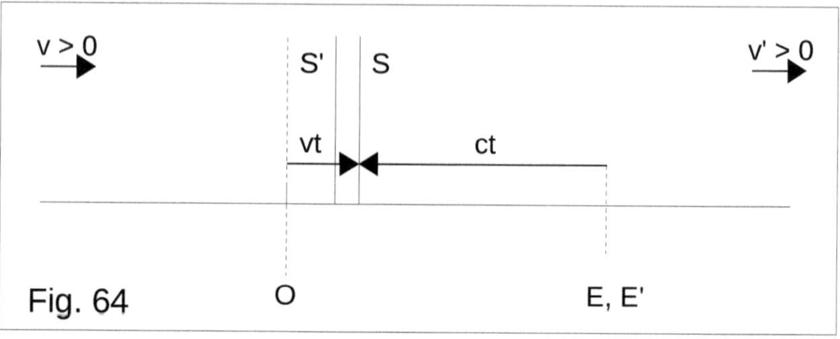

Fig. 64

We determine the coordinates of E.

$x = ct,\ d = ct + vt,\ d = t(c + v),\ t = d/(c + v) \rightarrow$

$\boldsymbol{E = (x, t) = (cd/(c + v),\ \ d/(c + v)).}$

The light signal continues and will reach S'. Meanwhile, S and S' move a bit to the right.

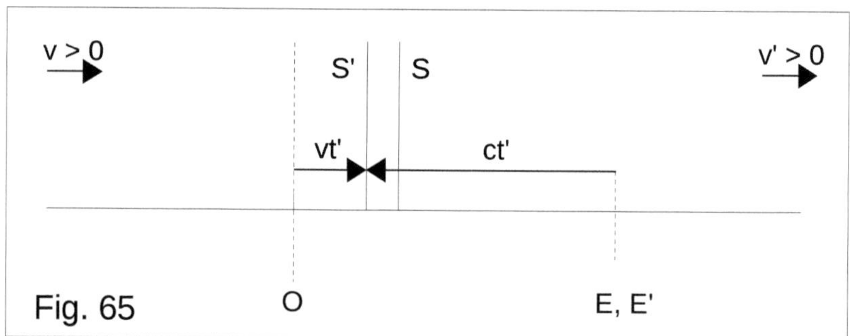

Fig. 65

We determine the coordinates of E'.

$x' = ct',\ d = ct' + v't',\ d = t'(c + v'),\ t' = d/(c + v') \rightarrow$

$\boldsymbol{E' = (x', t') = (cd/(c + v'),\ \ d/(c + v')).}$

8. Inertial reference systems – light signal from the left

Thought Experiment L1

We consider two inertial reference systems, S and S'. At the beginning of the experiment, these two reference systems are located at the same point O on the x-axis. S moves to the **left** on the x-axis. S' is at **rest** relative to S, $v' = v$.
This means that S' also moves to the left at the same speed as S.

The point on the x-axis where the event occurs is denoted by the letter E, E' (event).
Note that it is ONE POINT but we will consider it from two different reference systems, S and S'.
Therefore, we will denote the coordinates of the event with:

$E = (x, t)$ *for reference system S*
$E' = (x', t')$ *for reference system S'*

At the beginning of the experiment we have
$t = 0, t' = 0$.

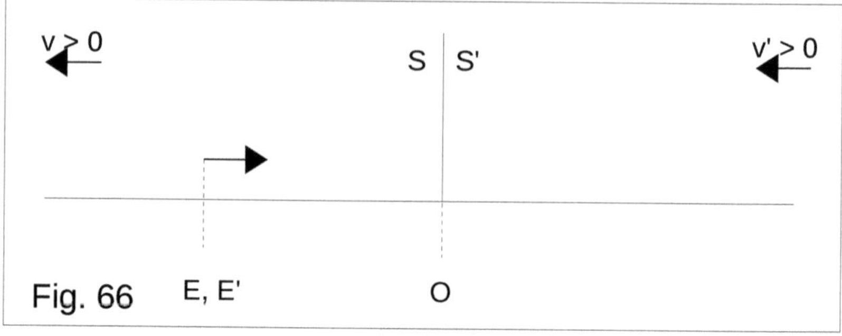

Fig. 66

At that moment, a light signal occurs on the x-axis. The light source is at a distance *d* from point O.

As the light signal moves towards S and S', S and S' move a short distance to the left. They move at the same speed.

The light signal will reach both reference systems simultaneously.

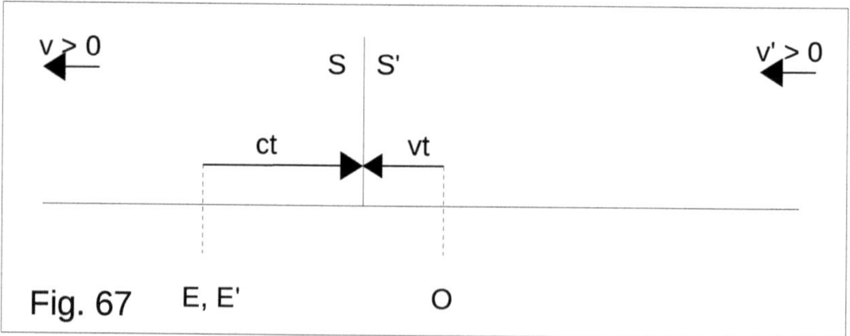

Fig. 67

We determine the coordinates of E.

$x = ct,\ d = ct + vt,\ d = t(c + v),\ t = d\,/\,(c + v) \rightarrow$
$\boldsymbol{E = (x, t) = (cd/(c + v),\ \ d/(c + v)).}$

It will be the same values for S'. In this case we have
$x' = x,\ t' = t,\ v' = v.$

$x' = ct',\ d = ct' + v't',\ d = t'(c + v'),\ t' = d\,/\,(c + v') \rightarrow$
$\boldsymbol{E' = (x', t') = (cd/(c + v'),\ \ d/(c + v')).}$

Thought Experiment L2

We consider **two** inertial reference systems, S and S'. At the beginning of the experiment, these two reference systems are located at the same point O on the x-axis. S moves to the **left** on the x-axis. S' moves to the **left** but at a lower speed than S, $v' < v$.

At the beginning of the experiment we have $t = 0, t' = 0$. At that moment a light signal occurs on the x-axis. The light source is at a distance d from point O.

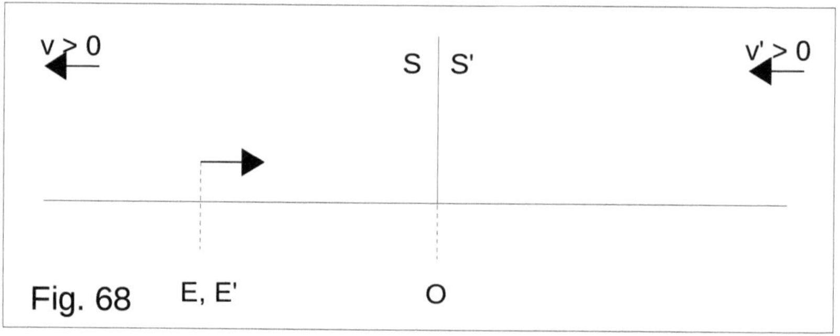

Fig. 68

As the light signal moves towards S and S', S and S' move to the left.

The light signal first reaches S.

We determine the coordinates of E.

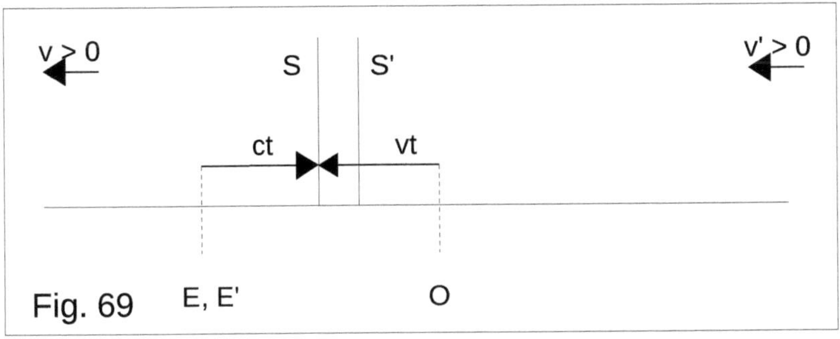

Fig. 69

$x = ct,\ d = ct + vt,\ d = t(c + v),\ t = d/(c + v) \rightarrow$
$\boldsymbol{E = (x, t) = (cd/(c + v),\ \ d/(c + v)).}$

The light signal continues and will reach S'. Meanwhile, S' moves more to the left.

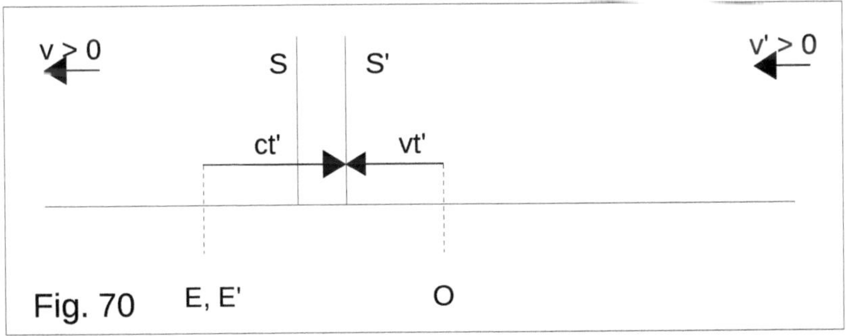

Fig. 70

We determine the coordinates of E'.

$x' = ct',\ d = ct' + v't',\ d = t'(c + v'),\ t' = d/(c + v') \rightarrow$
$\boldsymbol{E' = (x', t') = (cd/(c + v'),\ \ d/(c + v')).}$

Thought Experiment L3

We consider **two** inertial reference systems, S and S'. At the beginning of the experiment, these two reference systems are located at the same point O on the x-axis. S' is at **absolute rest** on the x-axis. S moves to the **left** relative to O.

At the beginning of the experiment we have $t = 0$, $t' = 0$. At that moment a light signal occurs on the x-axis. The light source is at a distance *d* from point O.

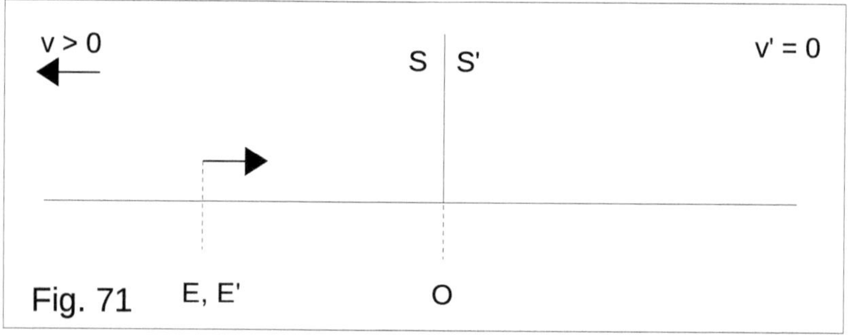

Fig. 71

While the light signal is moving towards S and S', S is moving a short distance to the left.

The light signal will first reach S.

We determine the coordinates of E.

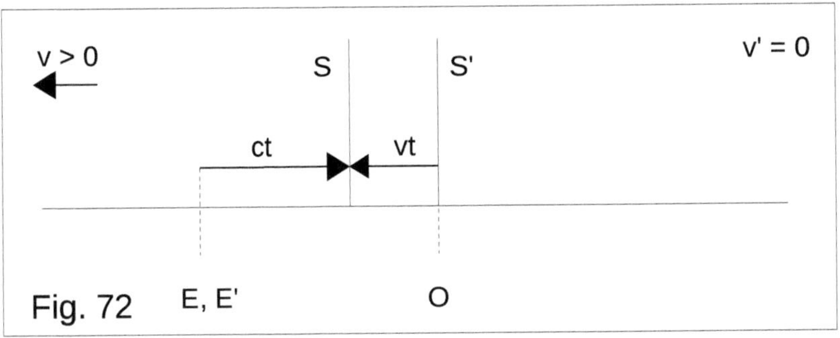

Fig. 72

$$x = ct,\ d = ct + vt,\ d = t(c + v),\ t = d/(c + v) \rightarrow$$
$$\boldsymbol{E = (x, t) = (cd/(c + v),\ \ d/(c + v))}.$$

The light signal continues and will reach S'.

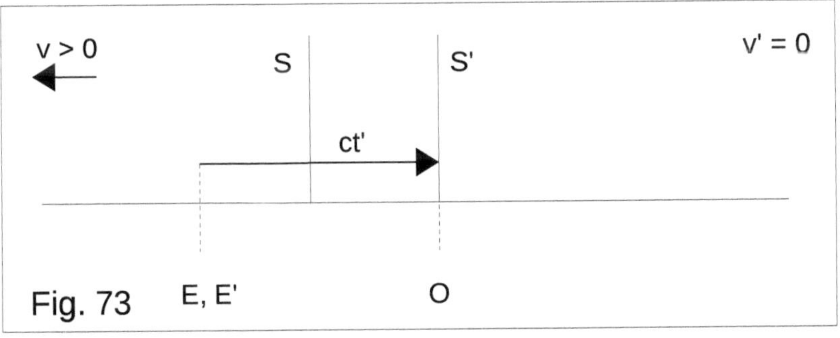

Fig. 73

We determine the coordinates of E'.

$$x' = ct',\ d = ct',\ t' = d/c \rightarrow$$
$$\boldsymbol{E' = (x', t') = (cd/c,\ \ d/c)}.$$

Thought Experiment L4

We consider **two** inertial reference systems, S and S'. At the beginning of the experiment, these two reference systems are located at the same point O on the x-axis. S moves to the **left** on the x-axis. S' moves to the **right** relative to O.

At the beginning of the experiment we have $t = 0, t' = 0$. At that moment a light signal occurs on the x-axis. The light source is at a distance d from point O.

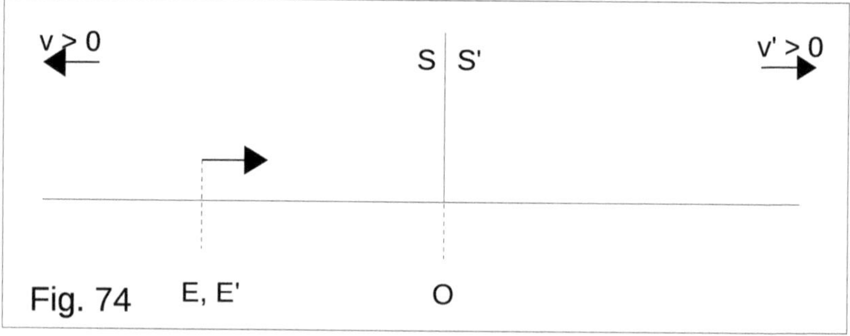

Fig. 74

As the light signal moves towards S and S', S moves to the left, S' to the right.

The light signal first reaches S.

We determine the coordinates of E.

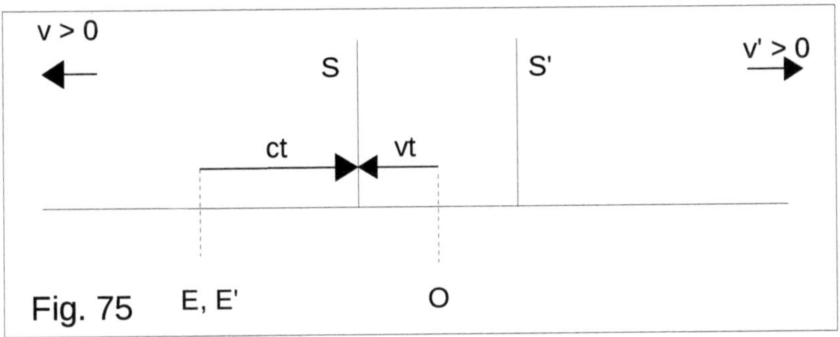

Fig. 75

$$x = ct,\ d = ct + vt,\ d = t(c + v),\ t = d/(c + v) \rightarrow$$
$$\boldsymbol{E = (x, t) = (cd/(c + v),\ \ d/(c + v)).}$$

The light signal continues and will reach S'. Meanwhile, S' moves more to the right.

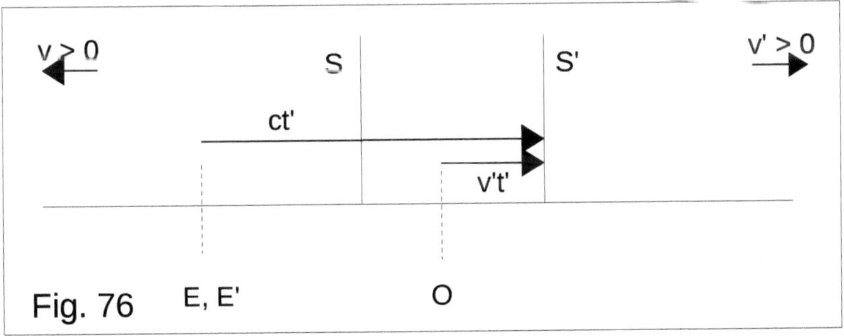

Fig. 76

We determine the coordinates of E'.

$$x' = ct',\ d = ct' - v't',\ d = t'(c - v'),\ t' = d/(c - v') \rightarrow$$
$$\boldsymbol{E' = (x', t') = (cd/(c - v'),\ \ d/(c - v')).}$$

Thought Experiment L5

We consider **two** inertial reference systems, S and S'. At the beginning of the experiment, these two reference systems are located at the same point O on the x-axis. S moves to the **left** on the x-axis. S' moves to the **left** but at a higher speed than S, $v' > v$.

At the beginning of the experiment we have $t = 0, t' = 0$. At that moment a light signal occurs on the x-axis. The light source is at a distance d from point O.

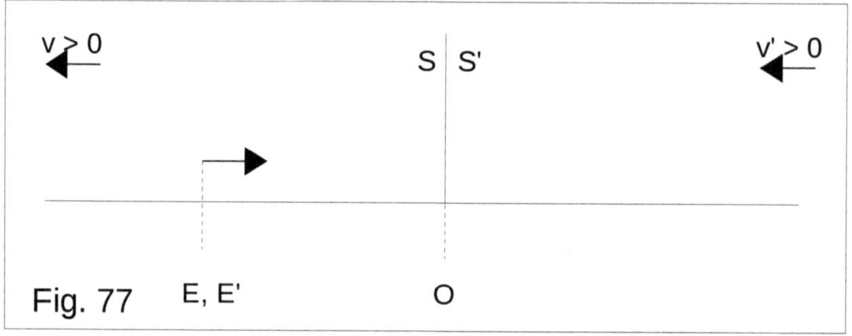

Fig. 77

As the light signal moves towards S and S', S and S' move to the left.

The light signal first reaches S'.

We determine the coordinates of E'.

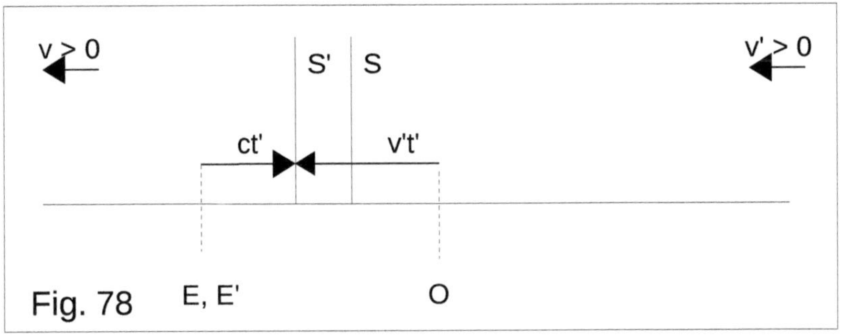

Fig. 78

$x' = ct'$, $d = ct' + v't'$, $d = t'(c + v')$, $t' = d/(c + v')$ →
$\boldsymbol{E' = (x', t') = (cd/(c + v'),\ d/(c + v'))}$.

The light signal continues and will reach S. Meanwhile, S moves more to the left.

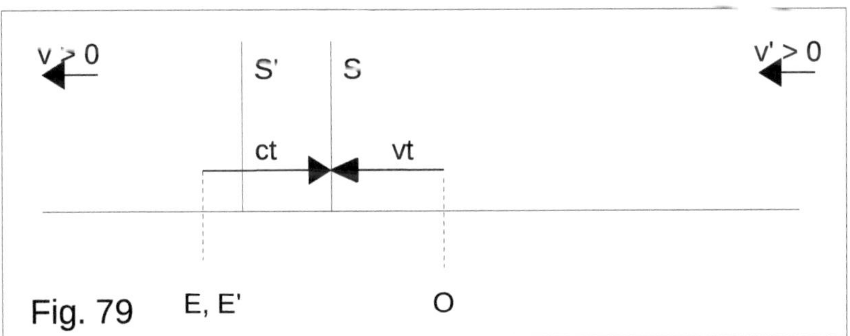

Fig. 79

We determine the coordinates of E.

$x = ct$, $d = ct + vt$, $d = t(c + v)$, $t = d/(c + v)$ →
$\boldsymbol{E = (x, t) = (cd/(c + v),\ d/(c + v))}$.

Thought Experiment L6

We consider **two** inertial reference systems, S and S'. At the beginning of the experiment, these two reference systems are located at the same point O on the x-axis. S is at **absolute rest** on the x-axis. S' moves to the **left** relative to O.

At the beginning of the experiment we have $t = 0, t' = 0$. At that moment a light signal occurs on the x-axis. The light source is at a distance d from point O.

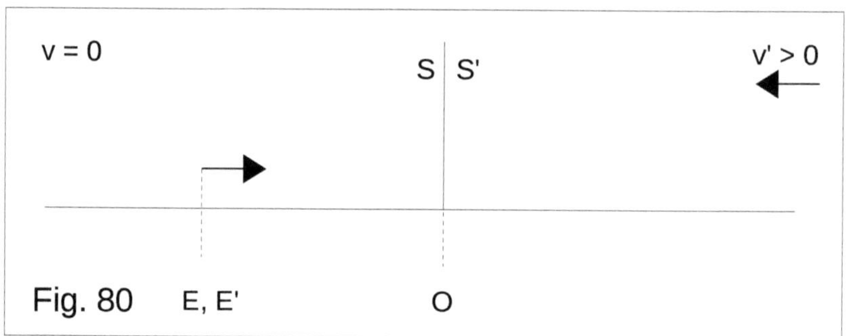

Fig. 80

While the light signal is moving towards S and S', S' is moving a short distance to the left.

The light signal will first reach S'.

We determine the coordinates of E'.

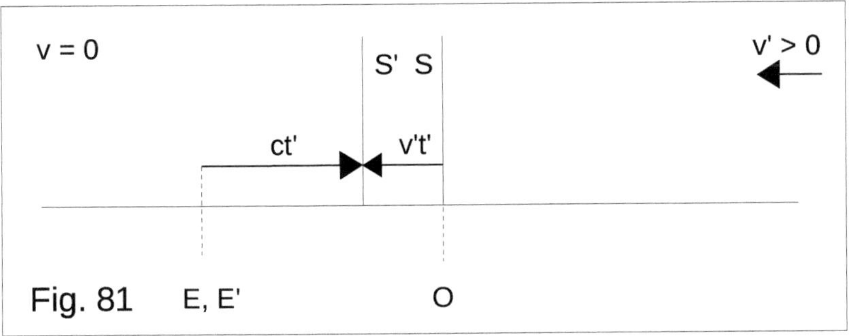

Fig. 81

$x' = ct',\ d = ct' + v't',\ d = t'(c + v'),\ t' = d/(c + v') \rightarrow$
$\boldsymbol{E' = (x', t') = (dc/(c + v'),\ \ d/(c + v'))}.$

The light signal continues and will reach S.

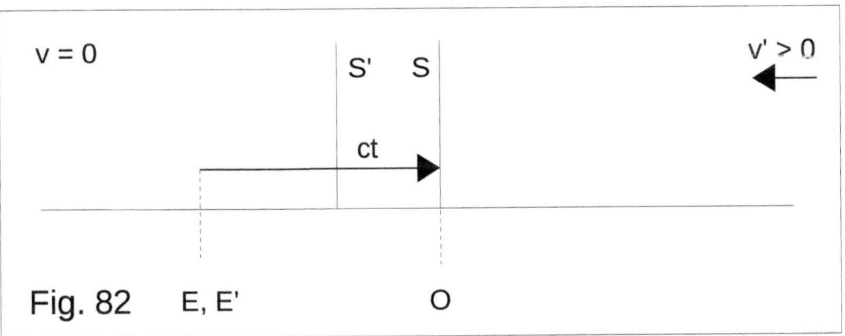

Fig. 82

We determine the coordinates of E.

$x = ct,\ d = ct,\ t = d/c \rightarrow$
$\boldsymbol{E = (x, t) = (cd/c,\ \ d/c)}.$

Thought Experiment L7

We consider **two** inertial reference systems, S and S'. At the beginning of the experiment, these two reference systems are located at the same point O on the x-axis. S moves to the **right** on the x-axis. S' moves to the **left** relative to O.

At the beginning of the experiment we have $t = 0, t' = 0$. At that moment a light signal occurs on the x-axis. The light source is at a distance d from point O.

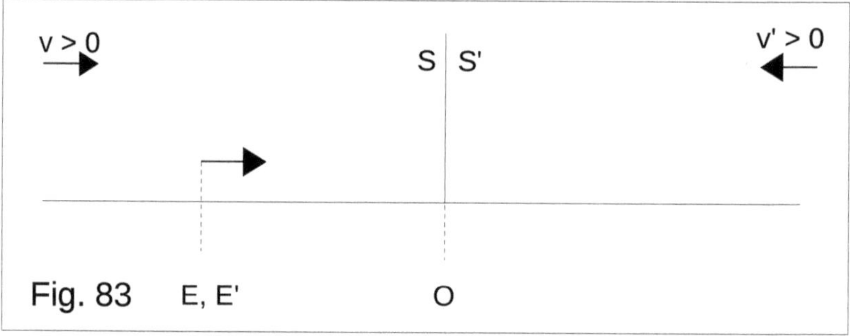

Fig. 83

As the light signal moves towards S and S', S moves to the right, S' to the left.

The light signal first reaches S'.

We determine the coordinates of E'.

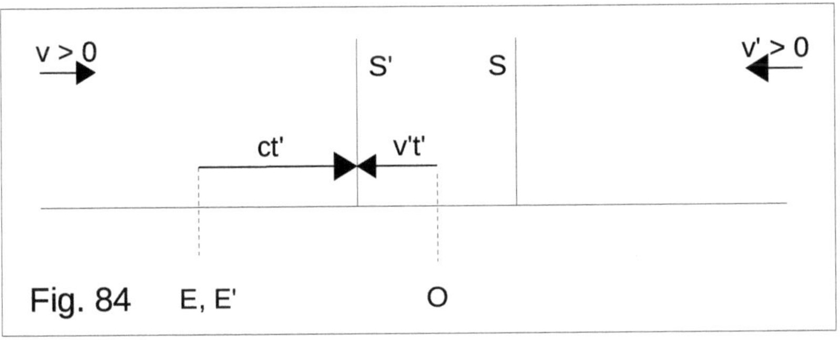

Fig. 84

$x' = ct', d = ct' + v't', d = t'(c + v'), t' = d/(c + v') \rightarrow$
$\boldsymbol{E' = (x', t') = (cd/(c + v'),\ d/(c + v'))}$.

The light signal continues and will reach S. Meanwhile, S moves more to the right.

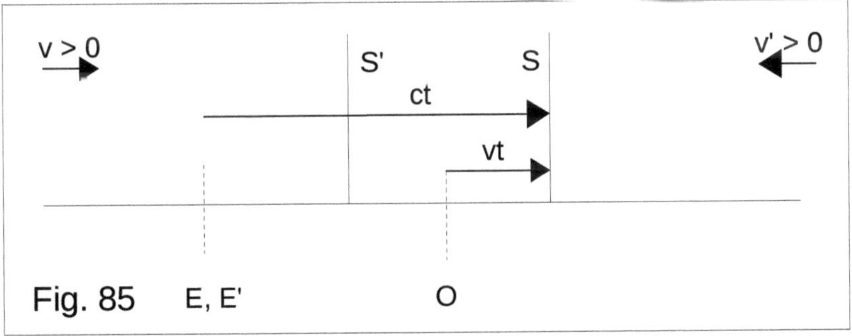

Fig. 85

We determine the coordinates of E.

$x = ct, d = ct - vt, d = t(c - v), t = d/(c - v) \rightarrow$
$\boldsymbol{E = (x, t) = (cd/(c - v),\ d/(c - v))}$.

Thought Experiment L8

We consider **two** inertial reference systems, S and S'. At the beginning of the experiment, these two reference systems are located at the same point O on the x-axis. These two reference systems are at rest relative to each other. And they are in **absolute rest** on the x-axis.

At the beginning of the experiment we have $t = 0$, $t' = 0$. At that moment a light signal occurs on the x-axis. The light source is at a distance *d* from point O.

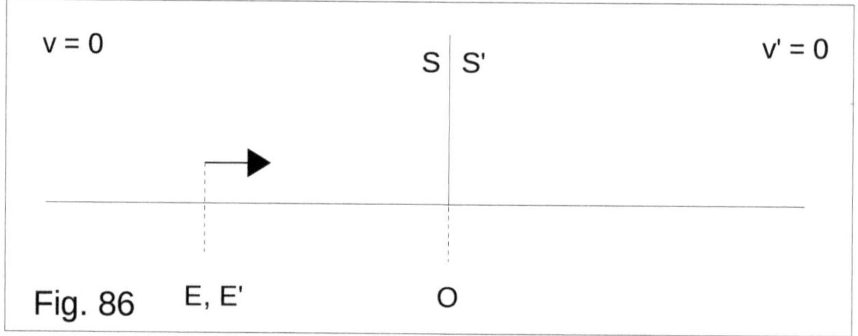

Fig. 86

When the light signal reaches the two reference systems, the times of the light signal arrival are registered.

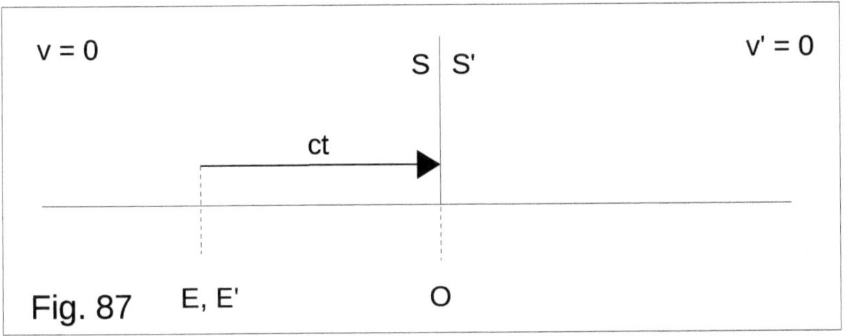

Fig. 87

We determine the coordinates of E.

$x = ct,\ d = ct,\ t = d\,/\,c \rightarrow$
$\boldsymbol{E = (x,\ t) = (cd/c,\ \ d/c).}$

In this case we have:

$x' = x$
$t' = t$
$v' = v = 0.$

We determine the coordinates of E'.

$x' = ct',\ d = ct',\ t' = d\,/\,c \rightarrow$
$\boldsymbol{E' = (x',\ t') = (cd/c,\ \ d/c).}$

Thought Experiment L9

We consider **two** inertial reference systems, S and S'. At the beginning of the experiment, these two reference systems are located at the same point O on the x-axis. S is at **absolute rest** on the x-axis. S' moves to the **right** relative to O.

At the beginning of the experiment we have $t = 0$, $t' = 0$. At that moment a light signal occurs on the x-axis. The light source is at a distance d from point O.

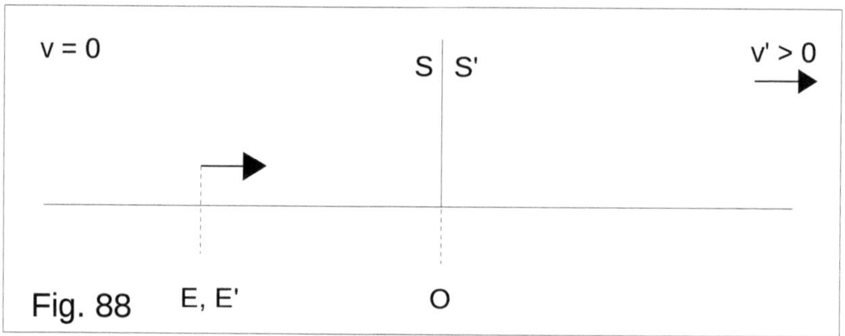

Fig. 88

While the light signal is moving towards S and S', S' is moving a short distance to the right.

The light signal will first reach S.

We determine the coordinates of E.

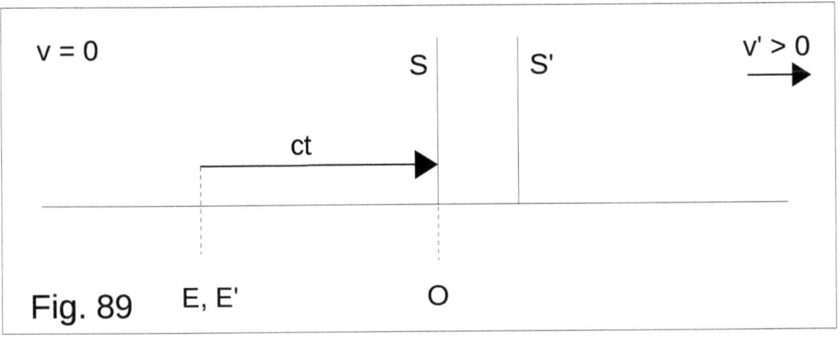

Fig. 89

$$x = ct,\ d = ct,\ t = d/c \rightarrow$$
$$\boldsymbol{E = (x, t) = (cd/c,\ \ d/c).}$$

The light signal continues and will reach S'. Meanwhile, S' moves a little further to the right.

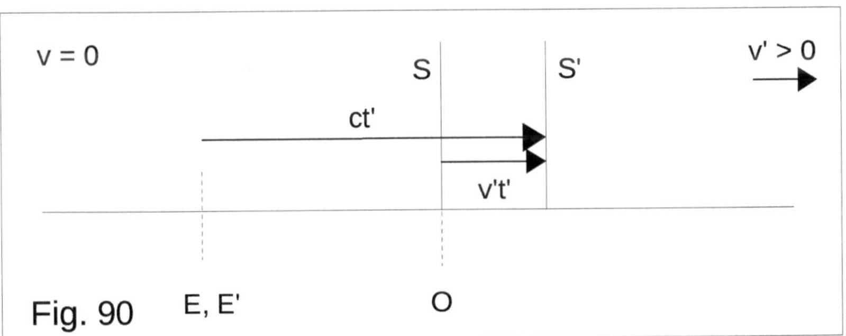

Fig. 90

We determine the coordinates of E'.

$$x' = ct',\ d = ct' - v't',\ d = t'(c - v'),\ t' = d/(c - v') \rightarrow$$
$$\boldsymbol{E' = (x', t') = (dc/(c - v'),\ d/(c - v')).}$$

Thought Experiment L10

We consider **two** inertial reference systems, S and S'. At the beginning of the experiment, these two reference systems are located at the same point O on the x-axis. S' is at **absolute rest** on the x-axis.
S moves to the **right** relative to O.

At the beginning of the experiment we have $t = 0, t' = 0$. At that moment a light signal occurs on the x-axis. The light source is at a distance d from point O.

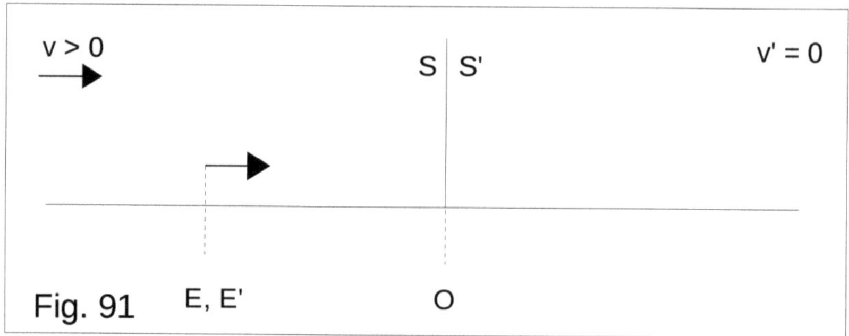

Fig. 91

While the light signal is moving towards S and S', S is moving a short distance to the right.

The light signal will first reach S'.

We determine the coordinates of E'.

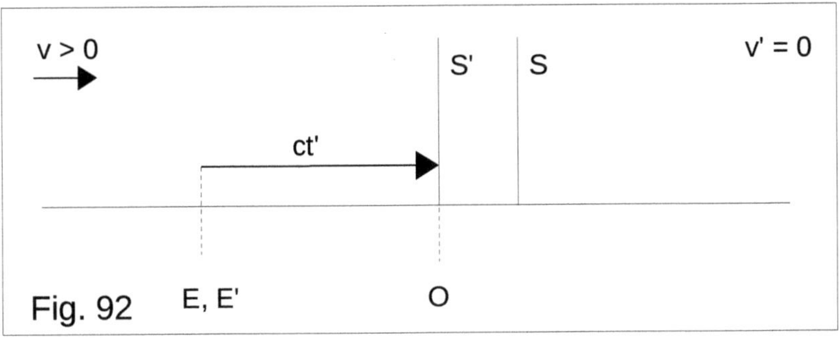

Fig. 92

$x' = ct', d = ct', t' = d / c \rightarrow$
$\boldsymbol{E' = (x', t') = (cd/c,\ d/c)}.$

The light signal continues and will reach S. Meanwhile, S moves a little further to the right.

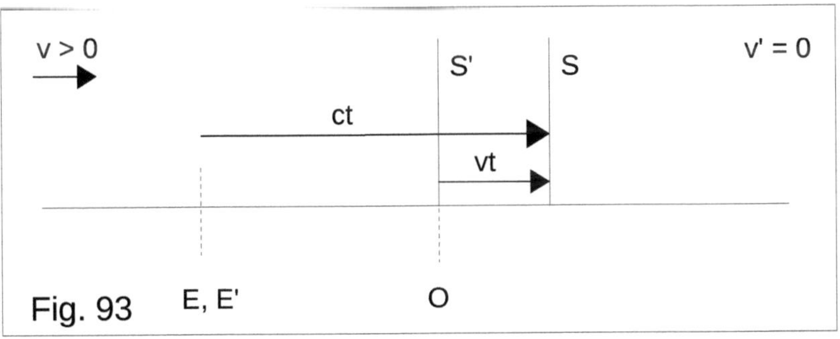

Fig. 93

We determine the coordinates of E.

$x = ct, d = ct - vt, d = t(c - v), t = d / (c - v) \rightarrow$
$\boldsymbol{E = (x, t) = (cd/(c - v), d/(c - v))}.$

Thought Experiment L11

We consider **two** inertial reference systems, S and S'. At the beginning of the experiment, these two reference systems are located at the same point O on the x-axis. S moves to the **right** on the x-axis. S' is at **rest** relative to S, $v' = v$.
This means that S' also moves to the right at the same speed as S.

At the beginning of the experiment we have
$t = 0$, $t' = 0$. At that moment a light signal occurs on the x-axis. The light source is at a distance d from point O.

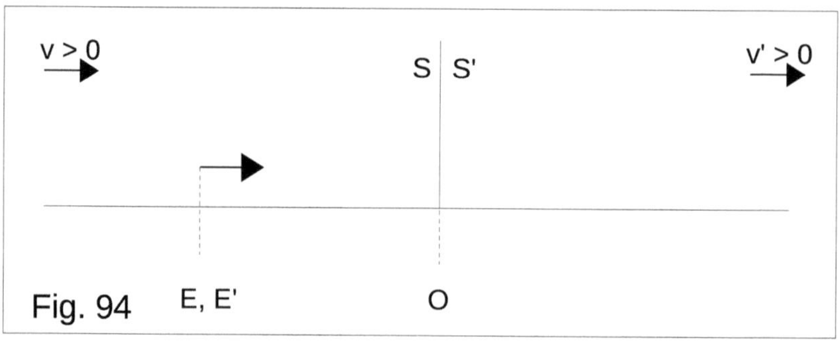

Fig. 94

As the light signal moves towards S and S', S and S' move a short distance to the right. They move at the same speed.

The light signal will reach both reference systems simultaneously.

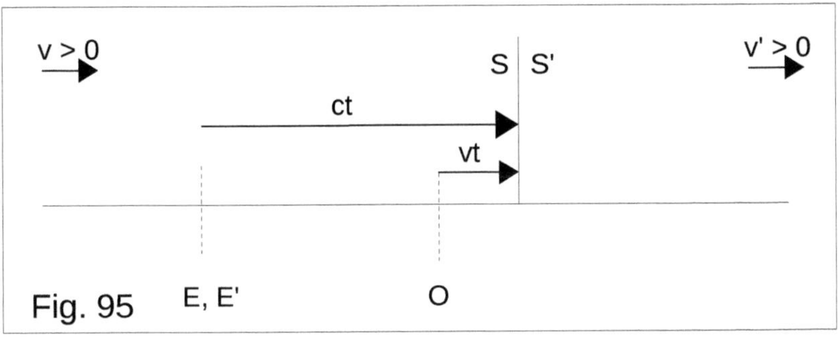

Fig. 95

We determine the coordinates of E.

$$x = ct,\ d = ct - vt,\ d = t(c - v),\ t = d/(c - v) \rightarrow$$
$$\boldsymbol{E = (x, t) = (cd/(c - v),\ \ d/(c - v)).}$$

It will be the same values for S'.
In this case we have

$$x' = x,\ t' = t,\ v' = v.$$

We determine the coordinates of E'.

$$x' = ct',\ d = ct' - v't' = t'(c - v'),\ t' = d/(c - v') \rightarrow$$
$$\boldsymbol{E' = (x', t') = (cd/(c - v'),\ \ d/(c - v')).}$$

Thought Experiment L12

We consider **two** inertial reference systems, S and S'. At the beginning of the experiment, these two reference systems are located at the same point O on the x-axis. S moves to the **right** on the x-axis. S' moves to the **right** but at a higher speed than S, $v' > v$.

At the beginning of the experiment we have $t = 0, t' = 0$. At that moment a light signal occurs on the x-axis. The light source is at a distance *d* from point O.

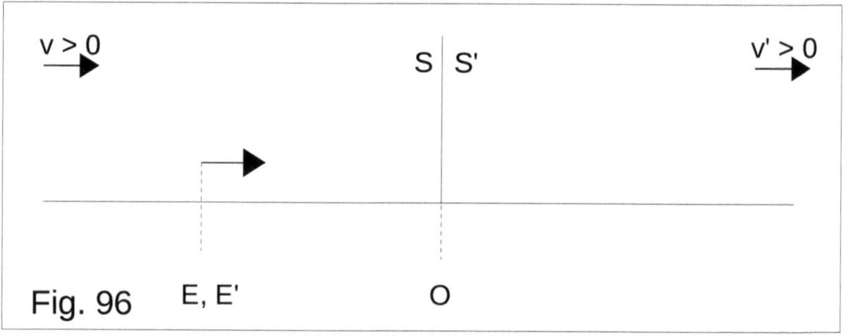

Fig. 96

As the light signal moves towards S and S', S and S' move a short distance to the right. S' moves faster than S.

The light signal will first reach S.

We determine the coordinates of E.

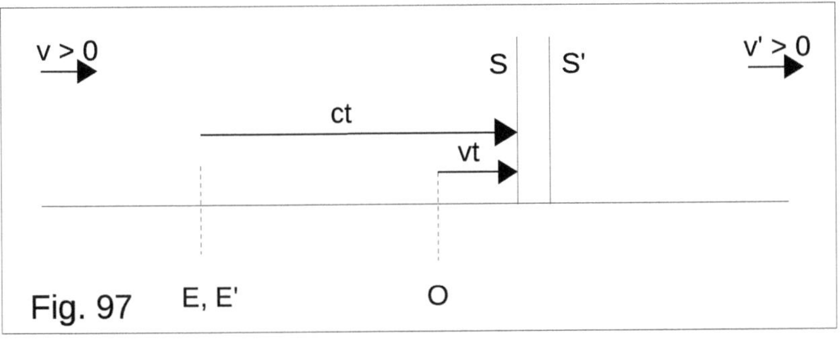

Fig. 97

$$x = ct,\ d = ct - vt,\ d = t(c - v),\ t = d/(c - v) \rightarrow$$
$$\boldsymbol{E = (x, t) = (cd/(c - v),\ \ d/(c - v)).}$$

The light signal continues and will reach S'. Meanwhile, S and S' move a little further to the right.

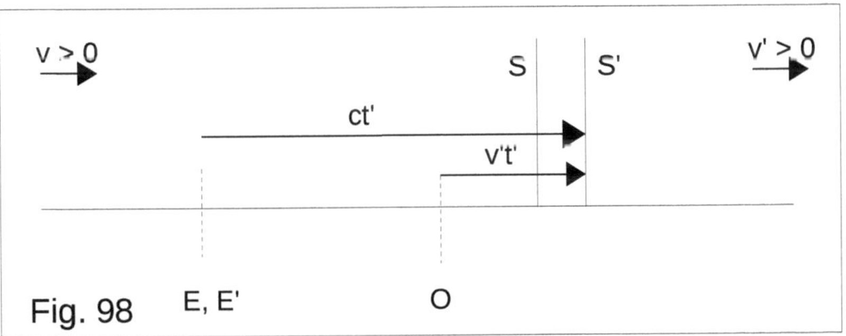

Fig. 98

We determine the coordinates of E'.

$$x' = ct',\ d = ct' - v't',\ d = t'(c - v'),\ t' = d/(c - v') \rightarrow$$
$$\boldsymbol{E' = (x', t') = (cd/(c - v'),\ \ d/(c - v')).}$$

Thought Experiment L13

We consider **two** inertial reference systems, S and S'. At the beginning of the experiment, these two reference systems are located at the same point O on the x-axis. S moves to the **right** on the x-axis. S' moves to the **right** but with less speed than S, $v' < v$.

At the beginning of the experiment we have $t = 0, t' = 0$. At that moment a light signal occurs on the x-axis. The light source is at a distance d from point O.

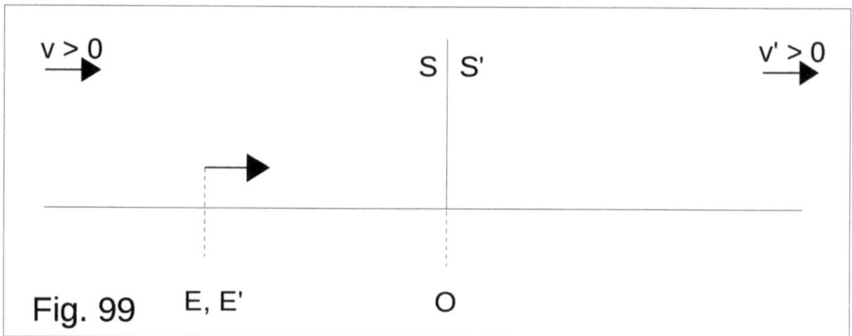

Fig. 99

As the light signal moves towards S and S', S and S' move a short distance to the right. S moves faster than S'.

The light signal will first reach S'.

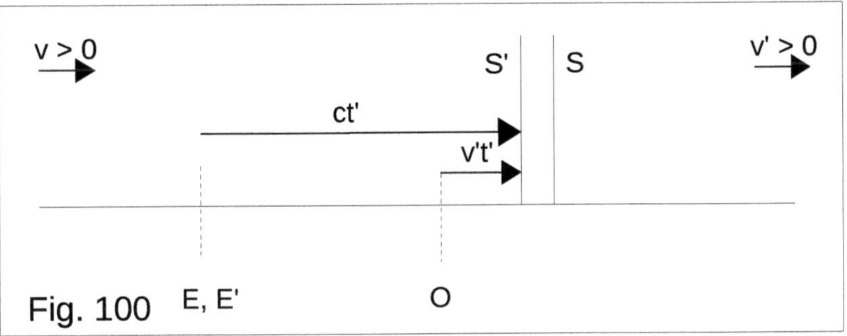

Fig. 100

We determine the coordinates of E'.

$x' = ct',\ d = ct' - v't',\ d = t'(c - v'),\ t' = d/(c - v') \ \rightarrow$

$\boldsymbol{E' = (x', t') = (cd/(c - v'),\ \ d/(c - v'))}.$

The light signal continues and will reach S. Meanwhile, S and S' move a bit to the right.

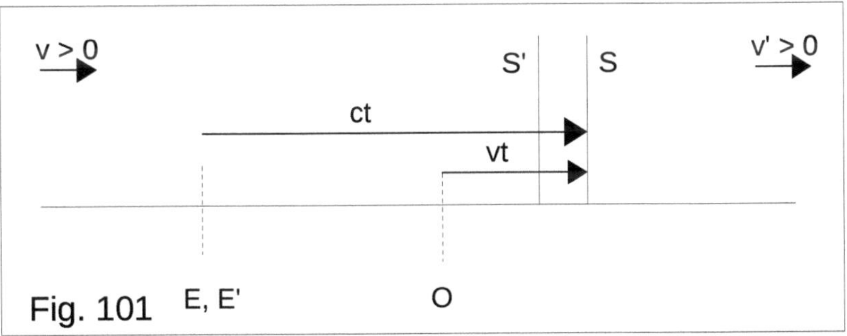

Fig. 101

We determine the coordinates of E.

$x = ct,\ d = ct - vt,\ d = t(c - v),\ t = d/(c - v) \rightarrow$

$\boldsymbol{E = (x, t) = (cd/(c - v),\ \ d/(c - v))}.$

9. Compilation of thought experiments R1-R13 and L1-L13

We compile the values for the coordinates of the events, as reference systems S and S' read them in the above 26 thought experiments.

R1

$E = (x, t) = (cd / (c - v), d / (c - v))$

$E' = (x', t') = (cd / (c - v'), d / (c - v'))$

$\rightarrow$

$x' = x(c - v) / (c - v'), t' = t(c - v) / (c - v')$

R2

$E' = (x', t') = (cd / (c - v'), d / (c - v'))$

$E = (x, t) = (cd / (c - v), d / (c - v))$

$\rightarrow$

$x' = x(c - v) / (c - v'), t' = t(c - v) / (c - v')$

R3

$E' = (x', t') = (cd / c, d / c)$

$E = (x, t) = (cd / (c - v), d / (c - v))$

$\rightarrow$

$x' = x(c - v) / (c \pm 0), t' = t(c - v) / (c \pm 0)$

R4

$E' = (x', t') = (cd/(c + v'), d/(c + v'))$

$E = (x, t) = (cd/(c - v), d/(c - v))$

$\rightarrow$

$x' = x(c - v)/(c + v'), t' = t(c - v)/(c + v')$

R5

$E = (x, t) = (cd/(c - v), d/(c - v))$

$E' = (x', t') = (cd/(c - v'), d/(c - v'))$

$\rightarrow$

$x' = x(c - v)/(c - v'), t' = t(c - v)/(c - v')$

R6

$E = (x, t) = (cd/c, d/c)$

$E' = (x', t') = (cd/(c - v'), d/(c - v'))$

$\rightarrow$

$x' = x(c \pm 0)/(c - v'), t' = t(c \pm 0)/(c - v')$

R7

$E = (x, t) = (cd/(c + v), d/(c + v))$

$E' = (x', t') = (cd/(c - v'), d/(c - v'))$

$\rightarrow$

$x' = x(c + v)/(c - v'), t' = t(c + v)/(c - v')$

R8

$E = (x, t) = (cd / c, d / c)$

$E' = (x', t') = (cd / c, d / c)$

$\rightarrow$

$x' = x(c \pm 0) / (c \pm 0), t' = t(c \pm 0) / (c \pm 0)$

R9

$E' = (x', t') = (cd / (c + v'), d / (c + v'))$

$E = (x, t) = (cd / c, d / c)$

$\rightarrow$

$x' = x(c \pm 0) / (c + v'), t' = t(c \pm 0) / (c + v')$

R10

$E = (x, t) = (cd / (c + v), d / (c + v))$

$E' = (x', t') = (cd / c, d / c)$

$\rightarrow$

$x' = x(c + v) / (c \pm 0), t' = t(c + v) / (c \pm 0)$

R11

$E = (x, t) = (cd / (c + v), d / (c + v))$

$E' = (x', t') = (cd / (c + v'), d / (c + v'))$

$\rightarrow$

$x' = x(c + v) / (c + v'), t' = t(c + v) / (c + v')$

R12

$E' = (x', t') = (cd/(c + v'), d/(c + v'))$

$E = (x, t) = (cd/(c + v), d/(c + v))$

$\rightarrow$

$x' = x(c + v)/(c + v'), t' = t(c + v)/(c + v')$

R13

$E = (x, t) = (cd/(c + v), d/(c + v))$

$E' = (x', t') = (cd/(c + v'), d/(c + v'))$

$\rightarrow$

$x' = x(c + v)/(c + v'), t' = t(c + v)/(c + v')$

L1

$E = (x, t) = (cd/(c + v), d/(c + v))$

$E' = (x', t') = (cd/(c + v'), d/(c + v'))$

$\rightarrow$

$x' = x(c + v)/(c + v'), t' = t(c + v)/(c + v')$

L2

$E = (x, t) = (cd/(c + v), d/(c + v))$

$E' = (x', t') = (cd/(c + v'), d/(c + v'))$

$\rightarrow$

$x' = x(c + v)/(c + v'), t' = t(c + v)/(c + v')$

L3

$E = (x, t) = (cd/(c + v), d/(c + v))$

$E' = (x', t') = (cd/c, d/c)$

$\rightarrow$

$x' = x(c + v)/(c \pm 0), t' = t(c + v)/(c \pm 0)$

L4

$E = (x, t) = (cd/(c + v), d/(c + v))$

$E' = (x', t') = (cd/(c - v'), d/(c - v'))$

$\rightarrow$

$x' = x(c + v)/(c - v'), t' = t(c + v)/(c - v')$

L5

$E' = (x', t') = (cd/(c + v'), d/(c + v'))$

$E = (x, t) = (cd/(c + v), d/(c + v))$

$\rightarrow$

$x' = x(c + v)/(c + v'), t' = t(c + v)/(c + v')$

L6

$E' = (x', t') = (cd/(c + v'), d/(c + v'))$

$E = (x, t) = (cd/c, d/c)$

$\rightarrow$

$x' = x(c \pm 0)/(c + v'), t' = t(c \pm 0)/(c + v')$

L7

$E' = (x', t') = (cd/(c + v'), d/(c + v'))$

$E = (x, t) = (cd/(c - v), d/(c - v))$

$\rightarrow$

$x' = x(c - v)/(c + v'), t' = t(c - v)/(c + v')$

L8

$E = (x, t) = (cd/c, d/c)$

$E' = (x', t') = (cd/c, d/c)$

$\rightarrow$

$x' = x(c \pm 0)/(c \pm 0), t' = t(c \pm 0)/(c \pm 0)$

L9

$E = (x, t) = (cd/c, d/c)$

$E' = (x', t') = (cd/(c - v'), d/(c - v'))$

$\rightarrow$

$x' = x(c \pm 0)/(c - v'), t' = t(c \pm 0)/(c - v')$

L10

$E' = (x', t') = (cd/c, d/c)$

$E = (x, t) = (cd/(c - v), d/(c - v))$

$\rightarrow$

$x' = x(c - v)/(c \pm 0), t' = t(c - v)/(c \pm 0)$

L11

$E = (x, t) = (cd/(c - v), d/(c - v))$

$E' = (x', t') = (cd/(c - v'), d/(c - v'))$

$\rightarrow$

$x' = x(c - v)/(c - v'), t' = t(c - v)/(c - v')$

L12

$E = (x, t) = (cd/(c - v), d/(c - v))$

$E' = (x', t') = (cd/(c - v'), d/(c - v'))$

$\rightarrow$

$x' = x(c - v)/(c - v'), t' = t(c - v)/(c - v')$

L13

$E' = (x', t') = (cd/(c - v'), d/(c - v'))$

$E = (x, t) = (cd/(c - v), d/(c - v))$

$\rightarrow$

$x' = x(c - v)/(c - v'), t' = t(c - v)/(c - v')$

Of all 2x2x13 = 52 different forms of coordinates, we identify 6 that are distinct:

$E = (x, t) = (cd / c, d / c)$
→ occurs 6 times

$E' = (x', t') = (cd / c, d / c)$
→ occurs 6 times

$E = (x, t) = (cd / (c + v), d / (c + v))$
→ occurs 10 times

$E' = (x', t') = (cd / (c + v'), d / (c + v'))$
→ occurs 10 times

$E = (x, t) = (cd / (c - v), d / (c - v))$
→ occurs 10 times

$E' = (x', t') = (cd / (c - v'), d / (c - v'))$
→ occurs 10 times

The way our experiments were designed, we have

$v >= 0, v' >= 0,$

where v, v' are absolute velocities for our two inertial reference systems.

Of all 26 different **transformations**, we identify 9 that are distinct:

$$x' = x(c + v) / (c + v'),\ t' = t(c + v) / (c + v')$$
$$x' = x(c + v) / (c \pm 0),\ t' = t(c + v) / (c \pm 0)$$
$$x' = x(c + v) / (c - v'),\ t' = t(c + v) / (c - v')$$

$$x' = x(c \pm 0) / (c + v'),\ t' = t(c \pm 0) / (c + v')$$
$$x' = x(c \pm 0) / (c \pm 0),\ t' = t(c \pm 0) / (c \pm 0)$$
$$x' = x(c \pm 0) / (c - v'),\ t' = t(c \pm 0) / (c - v')$$

$$x' = x(c - v) / (c + v'),\ t' = t(c - v) / (c + v')$$
$$x' = x(c - v) / (c \pm 0),\ t' = t(c - v) / (c \pm 0)$$
$$x' = x(c - v) / (c - v'),\ t' = t(c - v) / (c - v')$$

Is it possible to create a single formula that would apply to all these variants?

However, we see that there is always a factor that is equal for the transformation between x' and x and between t' and t.

For example

$$x' = x\mathbf{(c + v)/(c + v')},\ t' = t\mathbf{(c + v)/(c + v')}$$

We call this factor the **relativity factor**, ***rf***.
This factor is a function of v, v'.

We denote this function of relativity factor, ***rf (v, v')***.

All of the 9 transformations above can be expressed with a single formula.

$$x' = x*rf(v, v'),\ t' = t*rf(v, v').$$

What we also see from this analysis of the 26 experiments is that the transformation between x', x is **not dependent** on t, t'.
And also, the transformation between t', t is **not dependent** on x', x.

> **This reflects my thought and my belief that there is**
> **no physical connection**
> **between space and time.**

The above relativity factor *rf* has the following 9 different expressions:

$$rf(v, v') = (c + v)/(c + v'),\ v > 0,\ v' > 0$$
for R11, R12, R13, L1, L2, L5

$$rf(v, v') = (c + v)/(c \pm 0),\ v > 0,\ v' = 0$$
for R10, L3

$$rf(v, v') = (c + v)/(c - v'),\ v > 0,\ v' > 0$$
for R7, L4

$$rf(v, v') = (c \pm 0)/(c + v'),\ v = 0,\ v' > 0$$
for R9, L6

$$rf(v, v') = (c \pm 0)/(c \pm 0),\ v = 0,\ v' = 0$$
for R8, L8

$$rf(v, v') = (c \pm 0)(c - v'),\ v = 0,\ v' > 0$$
for R6, L9

$$rf(v, v') = (c - v)/(c + v'),\ v > 0,\ v' > 0$$
for R4, L7

$$rf(v, v') = (c - v)/(c \pm 0),\ v > 0,\ v' = 0$$
for R3, L10

$$rf(v, v') = (c - v)/(c - v'),\ v > 0,\ v' > 0$$
for R1, R2, R5, L11, L12, L13

What do these formulas say about our experiments?

The expression $(c + v)$ applies when S approaches the light source.
The expression $(c + v')$ applies when S' approaches the light source.

The expression *(c ± 0)* applies when S is at absolute rest relative to the light source on x-axis.
The expression *(c ± 0)* applies when S' is at absolute rest relative to the light source on x'-axis.

The expression *(c – v)* applies when S moves away from the light source.
The expression *(c – v')* applies when S' moves away from the light source.

If we denote the absolute velocity of the reference system with v_s then the expression

$(c \pm v_s)$

represent

the relative speed

between the reference system and the wavefront of the light signal!

10. A simple radar

We will describe a device that measures the car's speed. The device can transmit electromagnetic signals and read the time when the reflected signal returns to the device.

In order to determine the car's speed, we must send two signals and read the times for the reflected signals. The car moves towards the device.

We make a mathematical model of this physical phenomenon by entering a 2-dimensional coordinate system. On one axis we depict distances and on the other time.

The observer who is sending signals and who register time is at the origo of the coordinate system. We depict the car as a rectangle.

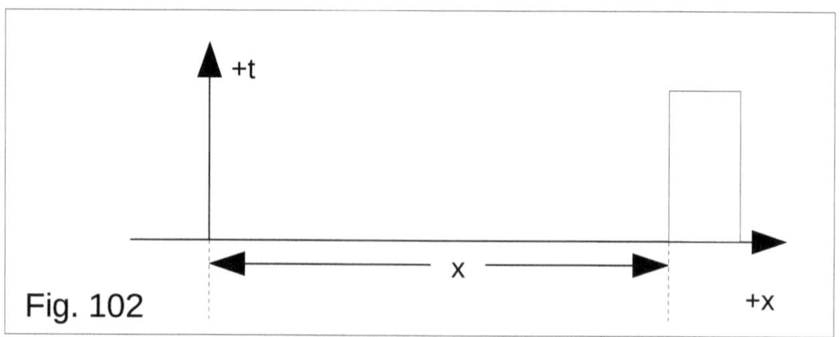

Fig. 102

When the first signal reaches the car, it is at a distance x from the origo.

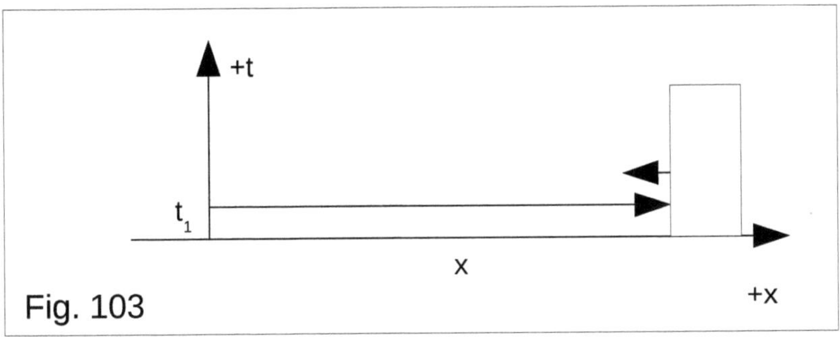

Fig. 103

When the first signal is transmitted, the device displays the time t_1.

The signal is reflected and returns to the observer. The time is read when the reflected signal reaches the device, t_2.

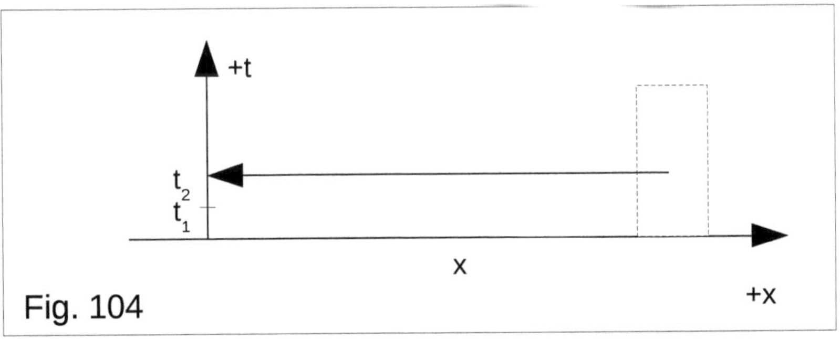

Fig. 104

The apparatus transmits the second signal at time t'_1. Meanwhile, the car has come closer to the observer.

When the signal reaches the car, it is at the distance x' from the device.
The signal is reflected and returns to the observer.

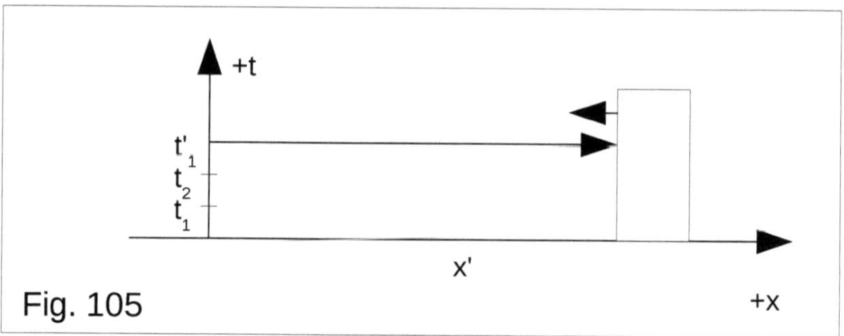

Fig. 105

When the second signal reaches the device, the time t'_2 is read.

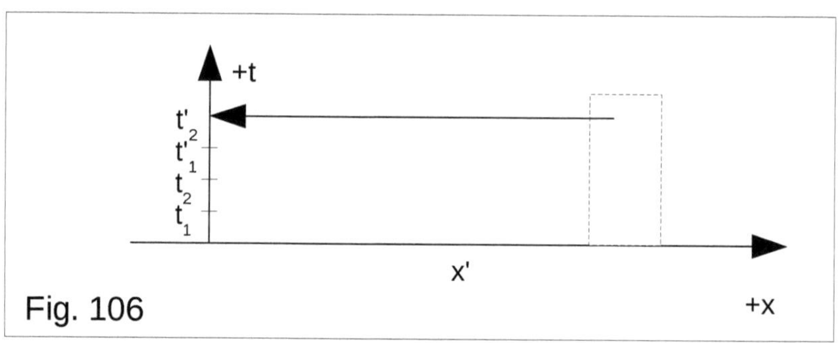

Fig. 106

Now the measurement is finished and we will now try to determine the speed v of the car.

Electromagnetic waves are moving at the speed of light, $c = 299\ 792\ 458\ m/s$.

The signal sent must go to the car and back to the device. Therefore, we will divide the time by *2*.

$$\Delta t = (t_2 - t_1)/2 = x/c$$
$$\Delta t' = (t'_2 - t'_1)/2 = x'/c$$

From here we can calculate x and x'.

$$x = (t_2 - t_1)c/2$$
$$x' = (t'_2 - t'_1)c/2$$

The car's speed becomes $v = (x - x')/T$.
How to determine the value of T?

We have two points on the x-axis in which the signal reaches the car, x and x'. These points, distance to the origo of the coordinate system, we have been able to calculate based on the read times for the signal.

Time T is the time interval during which the car passes the distance $x - x'$. Is it possible to determine, calculate, this time based on the 4 times we have read? It's probably not so easy!

Therefore, we must use the same mindset as we did in earlier parts of this book.

In models we build, we must create a relationship between

- the time and distance the signal pass and
- the time and distance the car pass.

When our experiment begins, the car is at point x_0.
The car approaches the observer, the device.
When the first signal is transmitted, at time t_1, the car is at point x_1.
This means that the distance $x_0 - x_1 = vt_1$.

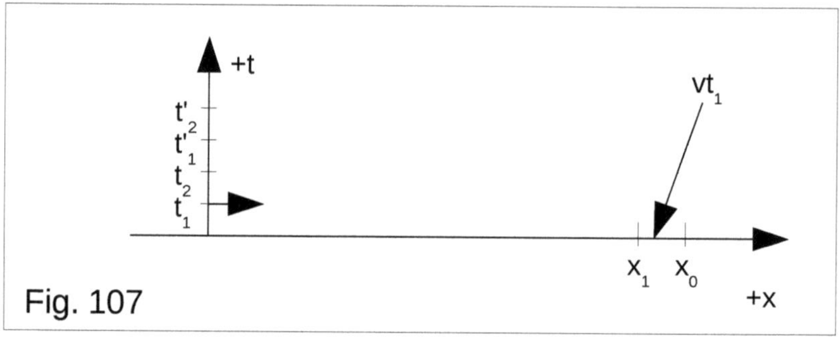

Fig. 107

The signal moves towards the car. We have decided before that the time the signal needs to reach the car is Δt.
During this time, the car moves to the left with a distance $v\Delta t$.

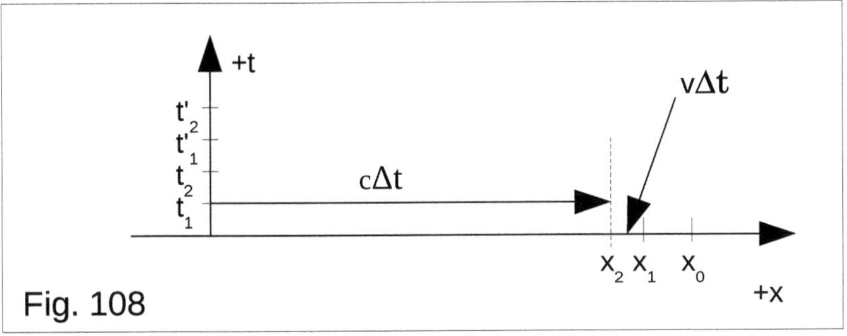

Fig. 108

The car continues to approach the observer, the apparatus, and when the second signal is transmitted, at time t'_1, the car has had time to pass the distance $v\ (t'_1 - t_2)$.

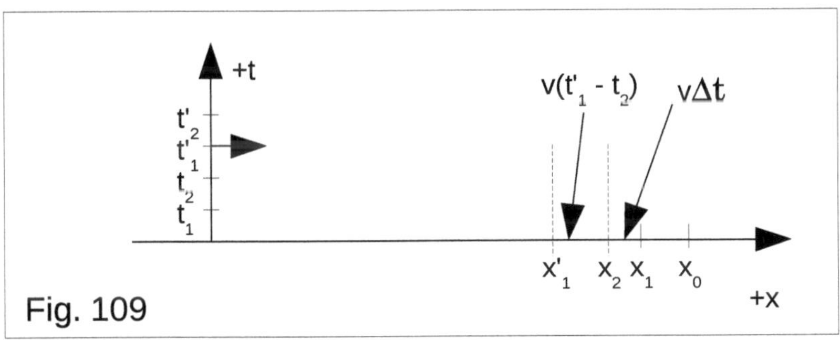

Fig. 109

The second signal moves towards the car. We have decided before that the time that the signal needs to reach the car is $\Delta t'$.
During this time, the car moves to the left with a distance $v\Delta t'$.

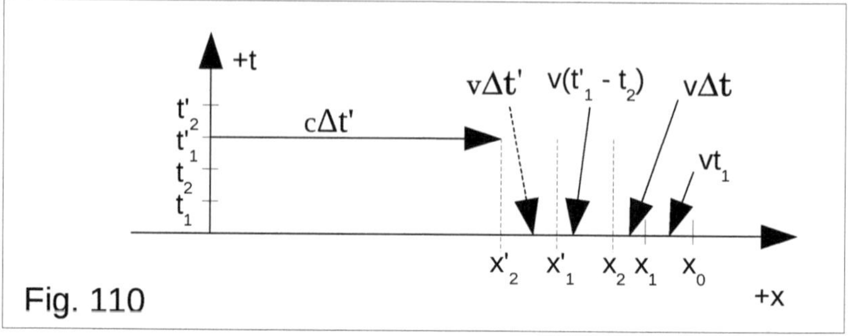

Fig. 110

Now we have all the components to be able to calculate the car's speed. We equate times and distances that apply to the first signal with times and distances that apply to the second signal.

We determine the distance between the origo of the coordinate system, say from point O to point x_0, the distance Ox_0, in two different ways.

$$c\Delta t + v\Delta t + vt'_1 =$$
$$c\Delta t' + v\Delta t' + v(t'_1 - t_2) + v\Delta t + vt'_1$$

The expression $v\Delta t + vt'_1$ is found in both terms

$$c\Delta t = c\Delta t' + v\Delta t' + v(t'_1 - t_2)$$

The time interval $t'_1 - t_2$ is known, it is about the time between the arrival time of the first signal back to the device and the time when the second signal is sent to the car. We can perform the experiment in such a way that the two times are equal, that the second signal is transmitted exactly when the first is read in the apparatus. Then $(t'_1 - t_2) = 0$.

We have left

$$c\Delta t = c\Delta t' + v\Delta t'$$
$$v\Delta t' = c\Delta t - c\Delta t'$$
$$\rightarrow$$
$$\boldsymbol{v = c(\Delta t - \Delta t')/\Delta t'}.$$

At the beginning we said that this chapter will be about a description of a simple radar. It was not that simple.
But to be able to determine the speed of a reference system (in our case it was a car) or for that matter the distance to this reference system, we need to read 4 times in our device.

We also see that we do not need to think about the wavelength, frequency, or Doppler effect, or any other physical properties of the electromagnetic signal we use.

We now verify our formula for calculating the velocity v. This is a theoretical verification.

Facts:
When the first signal reaches the car, it is at a distance of x = *100 m* from the observer.
When the second signal reaches the car, it is at a distance of x'= *90 m* from the observer.

From here we can calculate the time the signal needs to pass distances x and x'.

$$\Delta \boldsymbol{t} = \boldsymbol{x}/\boldsymbol{c}$$
$$\Delta \boldsymbol{t}' = \boldsymbol{x}'/\boldsymbol{c}$$
$$\rightarrow$$
$$\begin{aligned} v &= c(\Delta t - \Delta t')/\Delta t' \\ &= c(x/c - x'/c)/(x'/c) \\ &= c(x - x')/x' \\ &= c(100 - 90)/90 \\ &= c(10/90) \\ &= c/9 \end{aligned}$$

This means that for our example, the speed of the reference system is equal to one-ninth of the speed of light. It is not reasonable for a car.

We will try to set an example that is more reasonable.

Facts:
When the first signal reaches the car, it is at a

distance of $x = 100\ m$ from the observer.
The car's speed is $v = 100\ km/h$.

The question we ask is: what distance does the car travel while the two signals reach it?

$$\Delta t = x/c = 100\ \ m/c$$
$$v = 100\ km/h = ca\ 28\ m/s$$

So we have to calculate $\Delta t - \Delta t'$.
From the velocity formula above, we first calculate $\Delta t'$:

$$v = c(\Delta t - \Delta t')/\Delta t'.$$
$$v\Delta t' = c\Delta t - c\Delta t'$$
$$(c + v)\Delta t' = \ c\Delta t$$
$$\boldsymbol{\Delta t' = \Delta tc/(c + v)}$$
$$\rightarrow$$
$$\boldsymbol{\Delta t - \Delta t' = \Delta tv/(c + v)}$$

The distance we want to calculate becomes equal to
$(\Delta t - \Delta t')v$

$$= \Delta tv^2/(c + v)$$
$$= (x/c)v^2/(c + v)$$
$$= xv^2\ /(c(c + v))$$
$$= ca\ 8{,}7{*}10^{-13}\ m = ca\ 10{*}10^{-13}\ m = ca\ 10^{-12}\ m$$
$$= ca\ 10^{-9}\ millimeters$$

= ca 1 billionths of a millimeter

We see that this value is extremely small.
It is certainly difficult to read with great precision the four times we need in our experiment.

But this is a problem for the technicians who will build this simple radar.

Think about this example. How we determined time intervals and how we could calculate distance and speed.
Note that we did not use the special theory of relativity, we did not use the Lorentz factor.

We have used the current mathematics, simple ones. We have used the current logic.
And last but not least, we have used the current classical physics.

mathematics
logic
classical physics

11. Michelson-Morley experiment, 1887

Much has been written about this experiment. It is described as one of the most important and famous experiments in the history of physics.

But the result was not as expected.
It was so-called negative result. And this ultimately leads to the special theory of relativity.

In my analysis of this experiment, I show why it was so. I will show pictures of four different positions of the interferometer, show how the light rays move, make calculations on their passed distances.

In my analysis, I apply the principle that **the light moves independently of the source and the observer's motion**.

The pictures are not drawn to scale and I present only the most necessary elements to simplify as much as possible, so that one can see the most important.

In the first image, Fig. 111, we show the interferometer when the light beam is transmitted in the same direction as the apparatus moves.

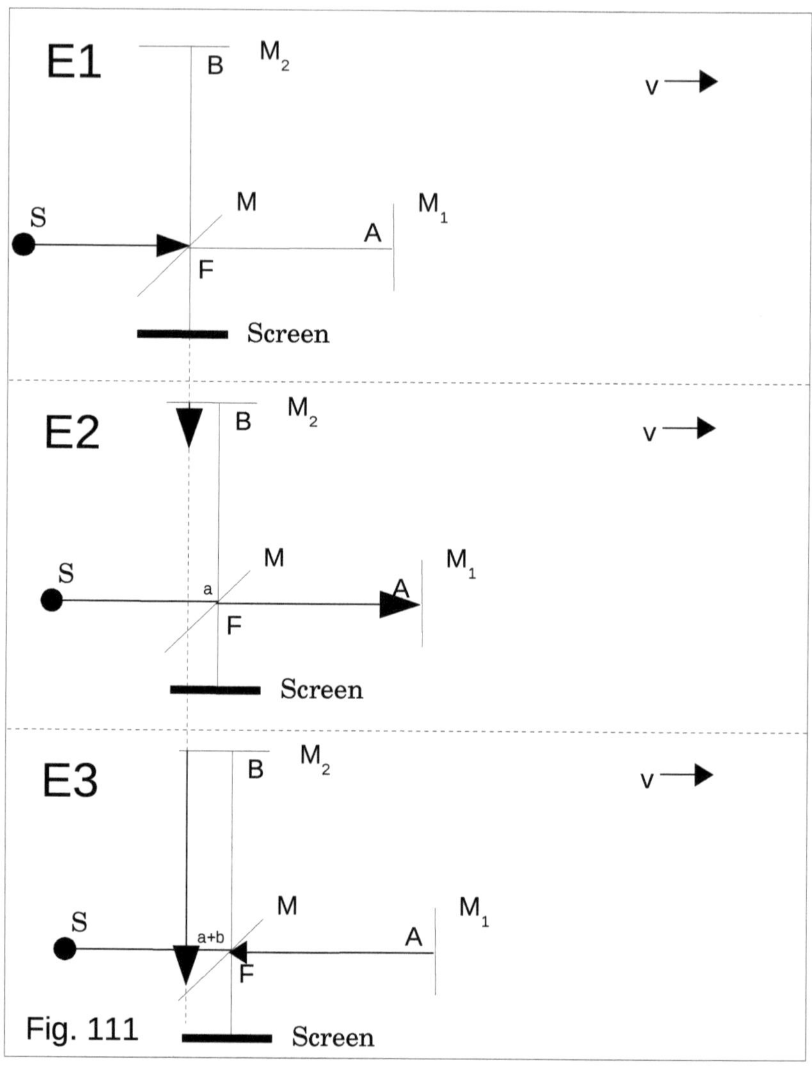

E1
B
M_2
v
S
M
M_1
A
F
Screen
E2
B
M_2
v
S
M
M_1
a
A
F
Screen
E3
B
M_2
v
S
M
M_1
a+b
A
F
Screen

Fig. 111

E1:
Interferometers arms: FA = FB = L (*length*).
A light beam is transmitted from S and is divided into two in F. The light beam S_1 continues straight ahead towards A (mirror M_1). S_2 is reflected at an angle of 90° and goes to B (mirror M_2).

E2:
When S_1 reaches A, it is reflected and goes back to F.
When S_2 reaches B, it is reflected and goes back to M.

While S_1 approaches A and reaches this point, the entire system moves with the distance a. During this time, S_2 pass the same distance $L+a$. $a/v = (L+a)/c$

E3:
While S_1 goes towards F and reaches this point, the entire system moves with the distance b. Then the S_1 pass the distance $L–b$. During this time, S_2 pass the same distance $L–b$. $b/v = (L–b)/c$

We now calculate the length of the distances that S1 and S2 pass.

$$a = Lv/(c–v)$$
$$b = Lv/(c+v)$$
$$a+b = 2Lcv/(c^2–v^2)$$
$$a–b = 2Lv^2/(c^2–v^2)$$

$$\text{Length}(S_1) = \text{Length}(S_2) = 2L + (a - b)$$

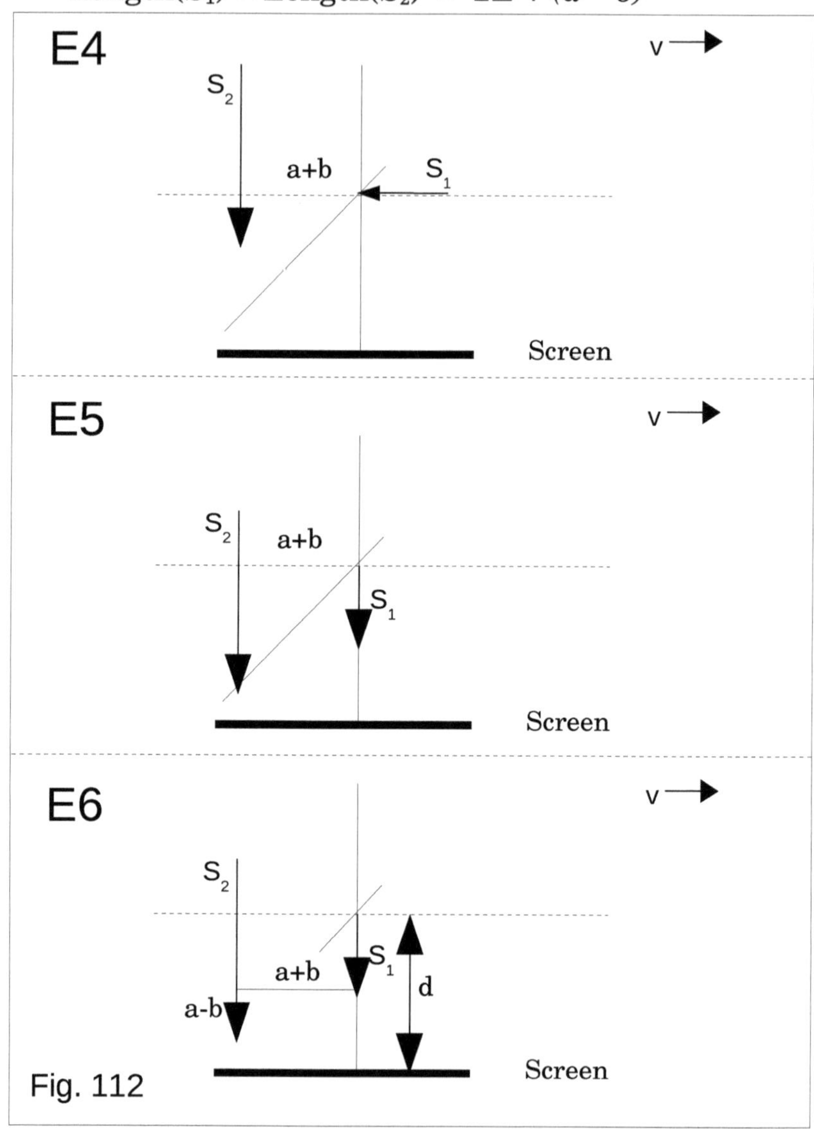

Fig. 112

In the figure Fig. 112 we show the moment when S_1 reaches the point F (mirror M) for the second time and then continues towards the screen.

Then the spatial distance between the two light rays becomes equal to $a + b$.
The light beam S_2 reaches earlier the screen, time difference then becomes $\Delta t = (a-b)/c$.

If we take the distance between the mirror M and the screen equal to s, the parameters of the interference pattern can be calculated. $y = s\lambda/d$

The size of the distance between two wave peaks of the interference pattern depends only on the distance s between the mirror M (point F) and the screen and the distance $d = a + b$ between the two light beams.

See picture E6 in Fig. 112.

We are now turning the device 90° counterclockwise. See Fig. 113.

We also make similar calculations here.

E7
A
M1
v
M2
M
Screen
B
F
S
E8
A
M1
v
M2
M
b
Screen
B
F
S
E9
A
M1
v
M2
M
Screen
B
F
S

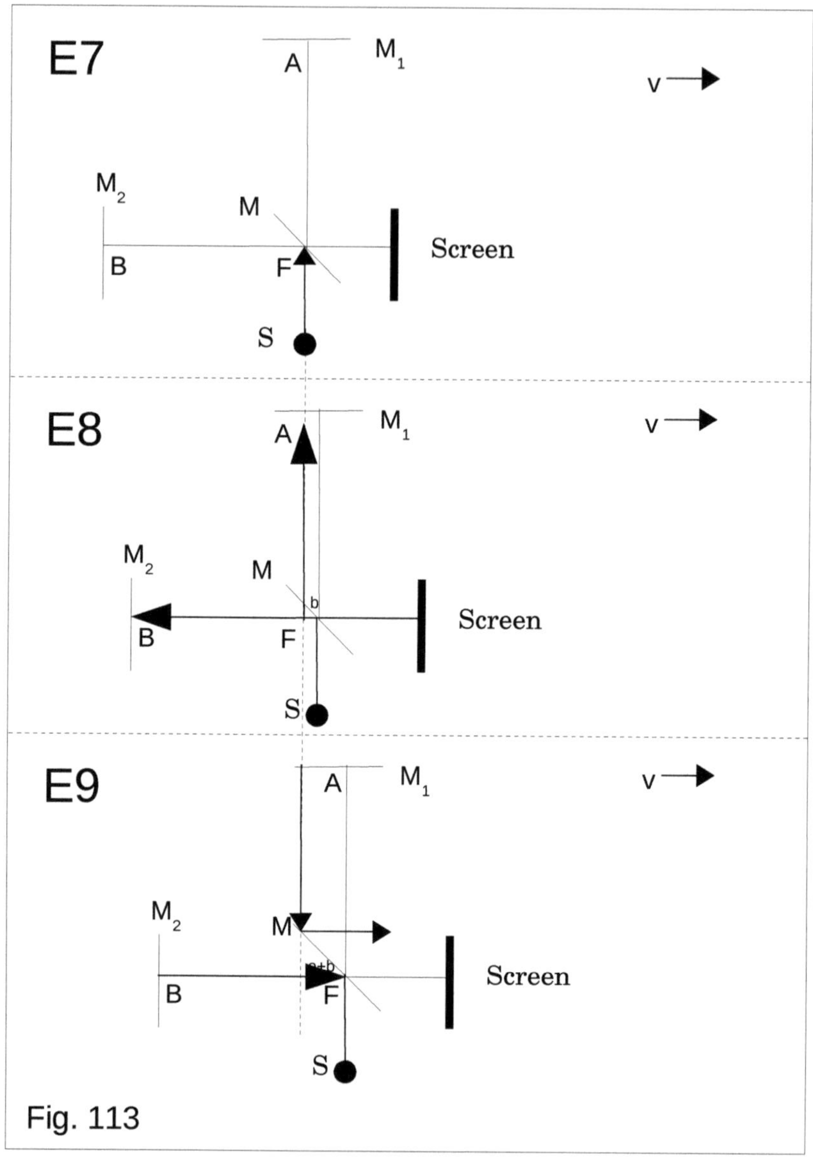

Fig. 113

E7:
Interferometers arms: FA = FB = L.
A light beam is transmitted from S and is divided into two in F. The light beam S_1 continues straight ahead towards A (mirror M_1). S_2 goes to B (mirror M_2).

E8:
When S_2 reaches B, it is reflected and goes back to F.
When S_1 reaches M_1, it is reflected and goes back to M.

As S_2 approaches B and reaches this point, the entire system moves with distance b. Then the S_2 pass the distance $L–b$. During this time, S_1 pass the same distance $L–b$.

E9:
While S_2 goes towards F and reaches this point, the entire system moves with a. Then the S_2 pass the distance $L + a$. During this time, S_1 pass the same distance $L + a$.

But we must calculate the moment when S_1 reaches the mirror M for the second time.
We analyze this in Fig. 114.

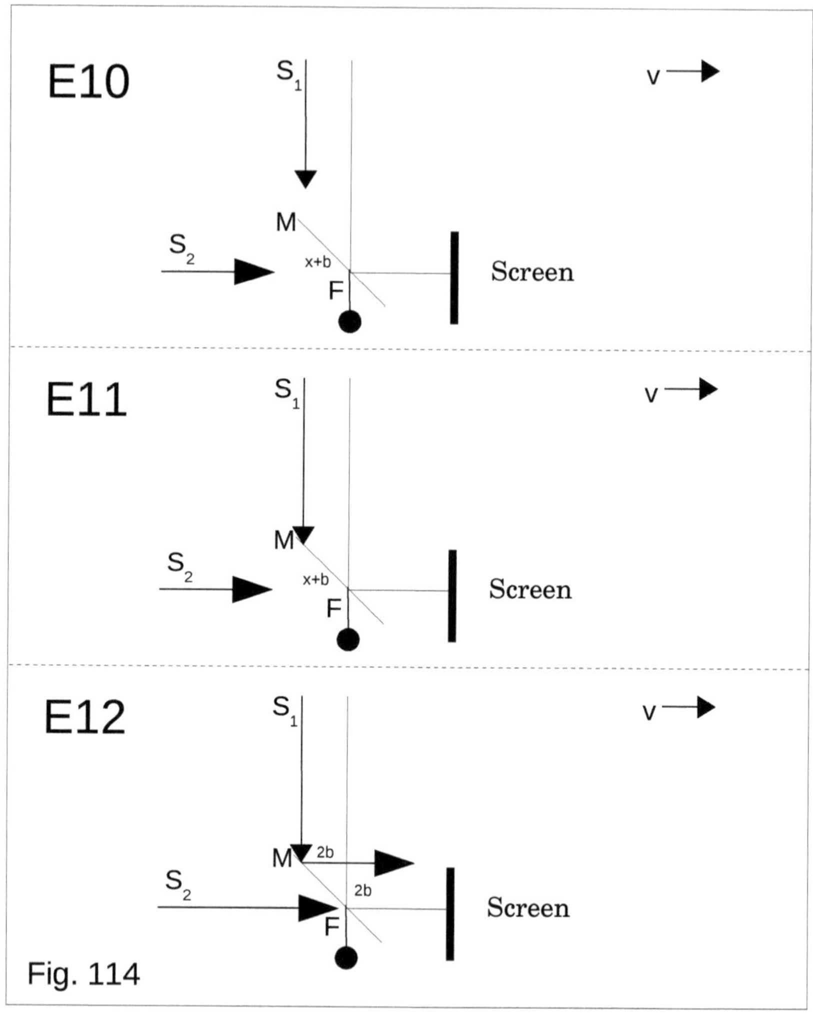

E10
S1
v
M
S2
x+b
F
Screen
E11
S1
v
M
S2
x+b
F
Screen
E12
S1
v
M
2b
S2
2b
F
Screen

Fig. 114

The calculation:

We say that the device pass the distance x at the moment the light beam S_1 reaches the mirror M for the second time.
The time for this is then x/v.

During this time, the light beam S_1 has traveled the distance $2L-x$.
Then we have:

$$x/v = (2L-x)/c$$
$$x/v = 2L/c-x/c$$
$$x/v+x/c = 2L/c$$
$$x(c+v)/cv = 2L/c$$
$$x = 2Lv/(c+v)$$

$$x = 2b$$

This means that the spatial distance between the two light beams is $d = 2b$.

We go on and turn the interferometer with 90° counterclockwise.

We show how the light rays move in Fig. 115.

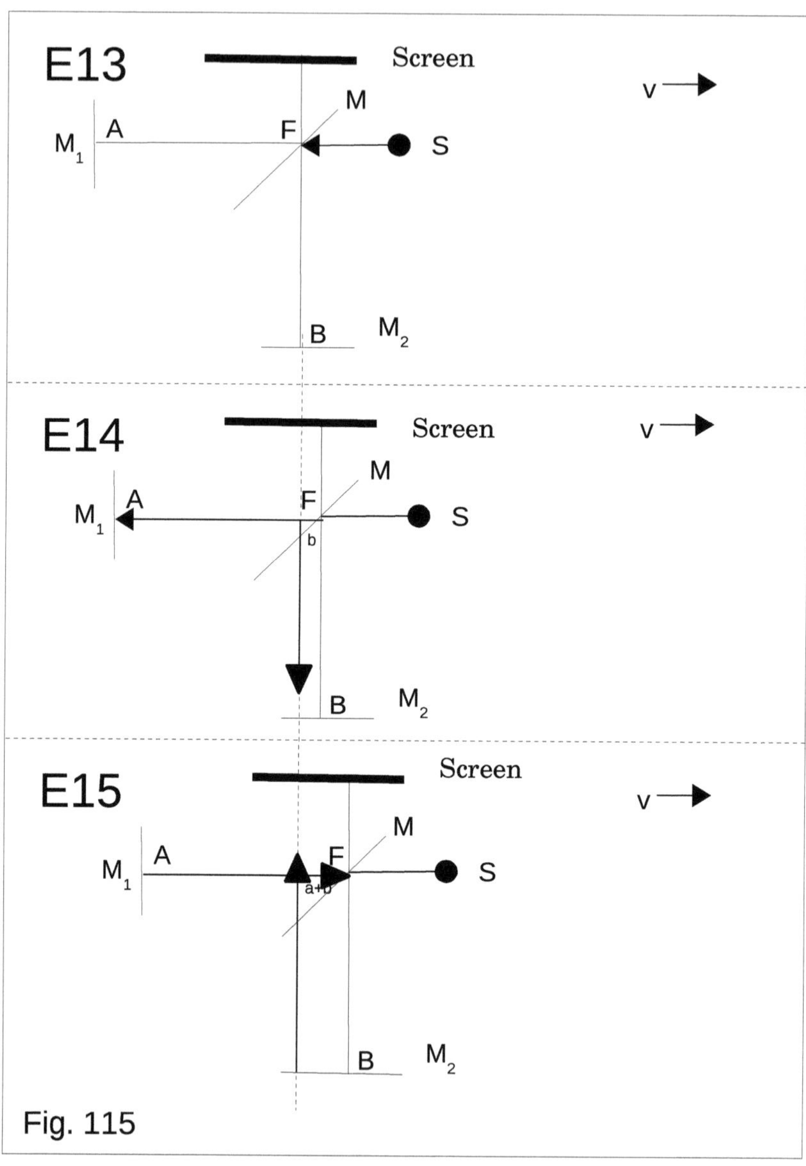

E13
Screen
v
M
A
F
M_1
S
B
M_2
E14
Screen
v
M
A
F
M_1
S
b
B
M_2
E15
Screen
v
M
A
F
M_1
S
a+b
B
M_2

Fig. 115

Interferometers arms: FA = FB = L.
A light beam is transmitted from S and is divided into two in F. The light beam S_1 continues straight ahead towards A (mirror M_1). S_2 goes to B (mirror M_2).

E14:
When S_1 reaches A, it is reflected and goes back to F. When S_2 reaches B, it is reflected and goes back to M.

While S_1 approaches A and reaches this point, the entire system moves by distance b. Then the S_1 pass the distance $L–b$. During this time, S_2 pass the same distance $L–b$.

E15:
While S_1 goes towards F and reaches this point, the whole system moves with a, S_1 pass the distance $L + a$. During this time, S_2 pass the same distance $L + a$.

This part of the experiment is similar to that of Fig. 111 in that the spatial distance between the two light beams when reaching the screen also becomes $d = a + b$.

Last part of the experiment. We turn the interferometer with another 90° counterclockwise. Now we have turned the interferometer with a full

270°. See picture Fig. 116.

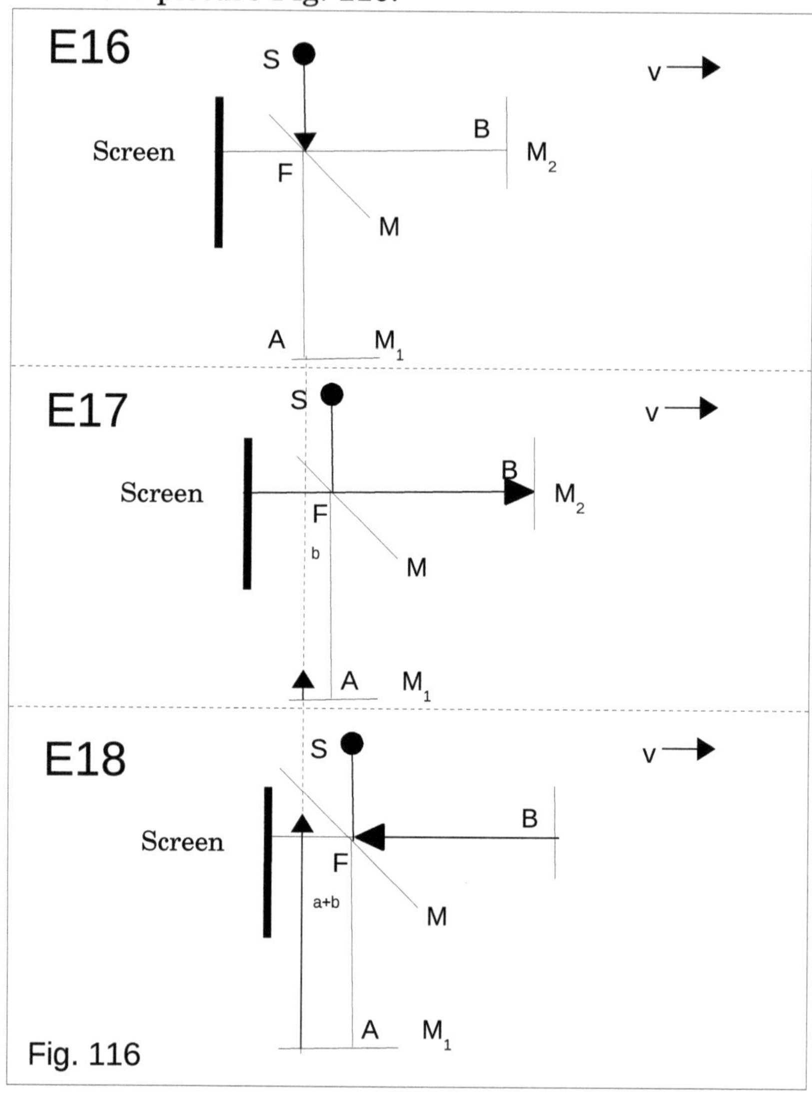

Fig. 116

E16:
Interferometers arms: $FA = FB = L$.
A light beam is transmitted from S and is divided into two in F. The light beam S_1 continues straight ahead towards A (mirror M_1). S_2 goes to B (mirror M_2).

E17:
When S_2 reaches B, it is reflected and goes back to F.
When S_1 reaches M_1, it is reflected and goes back to M.

While S_2 approaches B and reaches this point, the entire system moves by distance a. Then the S_2 pass the distance $L + a$. During this time, S_1 pass the same distance $L + a$.

E18:
While S_2 goes towards F and reaches this point, the whole system moves with b. Then the S_2 pass the distance $L–b$. During this time, S_1 pass the same distance $L–b$.

But we must calculate the moment when S_1 reaches the mirror M for the second time.

We analyze this in Fig. 117.

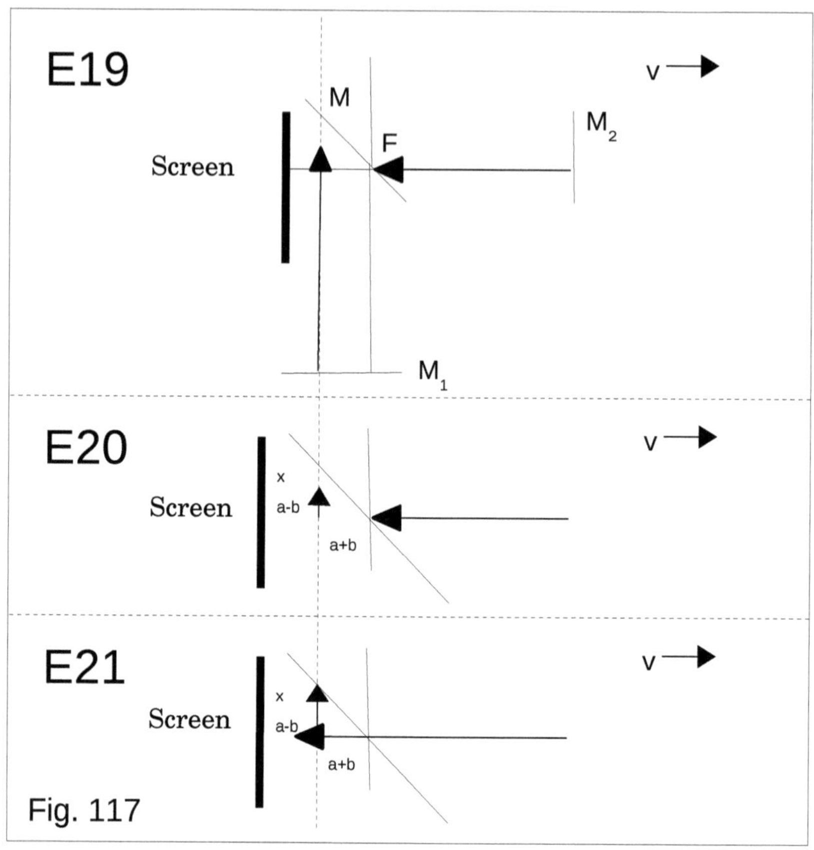

Fig. 117

When S_1 reaches mirror M for the second time, it has passed a distance of $2L + x$. The time then becomes $(2L + x)/c$ and is the same time as x/v.

$$x/v = (2L+x)/c$$
$$x = 2a$$

Then the spatial distance between the two light rays becomes equal to $2a$.

The spatial distance d between the two light beams reaching the screen is as follows:

Fig. 111-112	Exp. *1*	$d = a + b$
Fig. 113-114	Exp. *2*	$d = 2b$
Fig. 115	Exp. *3*	$d = a + b$
Fig. 116-117	Exp. *4*	$d = 2a$

We see that d reaches its maximum/minimum when the light beam is sent exactly at the right angle to the direction of motion of the apparatus. See Fig. 118.

Should we turn the middle mirror M by 180°
then $2b$ would change place with $2a$. And it is between these two part experiments when the interference pattern reaches the biggest difference!

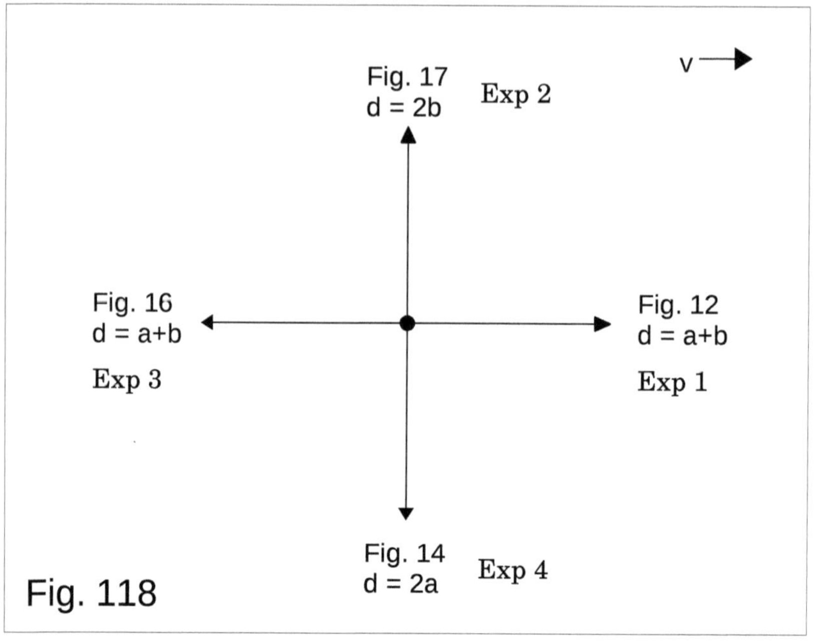

Fig. 118

But not even this is the whole truth! This applies if we direct the interferometer either along the same direction as the reference system moves or at right angles to it.

<u>When working with the light, it is extremely important to remember that you do NOT know the absolute speed of a reference system, nor its direction!</u>

The calculations we made before, we should redo and take into account that both interferometer arms can move sideways. See Fig. 119.

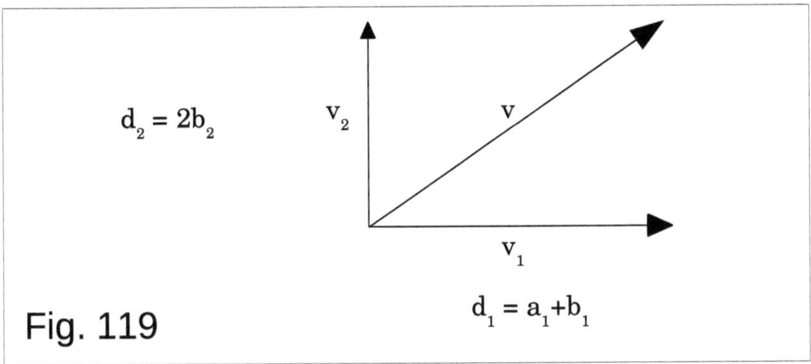

Fig. 119

Why? We do **not know** the exact direction of Earth's velocity nor its value.
We do not know the interferometer's absolute motion!

Because:
- The Earth has a motion around its own axis
- Earth moves around the sun
- The solar system moves around the center of the galaxy
- The galaxy moves towards the Great Attractor.
- And …

What is the result of all these speeds? How then does the interference pattern become?

In the table below, we show the calculation for *v = 18 km / s, 24 km / s, 30 km / s and 600 km / s.*

L	m	10	10	10	10
v	m/s	18000	24000	30000	600000
c	m/s	300000000	300000000	300000000	300000000
s	m	1	1	1	1
λ	m	0,0000006	0,0000006	0,0000006	0,0000006
a	m	0,0006000	0,0008001	0,0010001	0,0200401
b	m	0,0006000	0,0007999	0,0009999	0,0199601
a+b	m	0,0012000	0,0016000	0,0020000	0,0400002
a-b	m	0,0000001	0,0000001	0,0000002	0,0000800
d1=a+b	m	0,0012000	0,0016000	0,0020000	0,0400002
d2=2b	m	0,0011999	0,0015999	0,0019998	0,0399202
d3=a+b	m	0,0012000	0,0016000	0,0020000	0,0400002
d4=2a	m	0,0012001	0,0016001	0,0020002	0,0400802
y1=sλ/d1	m	0,0005000	0,0003750	0,0003000	0,0000150
y2=sλ/d2	m	0,0005000	0,0003750	0,0003000	0,0000150
y3=sλ/d3	m	0,0005000	0,0003750	0,0003000	0,0000150
y4=sλ/d4	m	0,0005000	0,0003750	0,0003000	0,0000150

In the table above we show some calculation values. *y1, y2, y3* and *y4* is the distance between two peaks in the interference pattern.

If you take *v = 30 km / s* as one did in the Michelson-Morley experiment, we see that the difference between the size of these distances is very small.

y1-y2	m	0,00000003	0,00000003	0,00000003	0,00000003
y2-y3	m	0,00000003	0,00000003	0,00000003	0,00000003
y3-y4	m	0,00000003	0,00000003	0,00000003	0,00000003
y4-y1	m	0,00000003	0,00000003	0,00000003	0,00000003

When turning the interferometer by 90° becomes $\Delta y = 0.000\ 000\ 030\ m$ which is *30 nanometers*. Could one read these small differences in 1887? Doubtful!

1 nm, a nanometer is equal to
0.000 000 001 meters and then above Δy is
0.000 000 030 meters (30 nm)

It is no wonder that one could not determine how it is with the Earth's velocity towards the ether!
It is no wonder that one made the wrong conclusion about the ether! It's no wonder that it was a negative result!

What is most strange in my calculations is that for a value of v you get approximately the same Δy between the different sub-moments, when the interferometer is turned by 90°. And keep in mind that I have used $v = 30\ km/s$ and $600\ km/s$. It's 20 times more in the second case! This means that Michelson interferometer was **not suitable for use for its intended purpose**, to measure the Earth's velocity in space!

In this analysis, I have applied calculations that apply to 2-slot interference. In the Michelson-Morley experiment, one uses time difference between the two light rays. Below we also do this analysis.

Fig. 111-112	Exp. *1*	$\Delta t = (a-b)/c$
Fig. 113-114	Exp. *2*	$\Delta t = 0$
Fig. 115	Exp. *3*	$\Delta t = (a-b)/c$
Fig. 116-117	Exp. *4*	$\Delta t = 0$

This is calculated in a similar way as we calculated the spatial distance between S_1 and S_2.

We calculate $N = \Delta t / T$ where $T = 1{,}83 * 10^{-15}$ s.

Fig. 111-112	Exp. *1*	$N = 0{,}364$ *(0,40)*
Fig. 113-114	Exp. *2*	$N = 0$
Fig. 115	Exp. *3*	$N = 0{,}364$ *(0,40)*
Fig. 116-117	Exp. *4*	$N = 0$

In Exp. 2 and 4, there is no time difference between the two light rays, which means that no interference pattern should be created! Did it?

We look at the pictures in the above four experiments that for the light rays S_1 and S_2 there is both a spatial and temporal 'distance'. How are the interference patterns formed in these cases?

12. My patent application: 1ANM

1) 1ANM, Description

Apparatus for measuring of the absolute velocity in space – 1ANM (One Arm, No Mirrors)

Background

[0001]
Historical Facts:
- 1864: James Clark Maxwell publishes *Dynamical Theory of the Electromagnetic Field.* Derive that light is an electromagnetic wave. The light should propagate in a medium in the same way that sound waves need air, water waves need water.
This medium was called ether, light-bearing ether.
- 1887: The Michelson-Morley experiment. The purpose of the experiment was to confirm the existence of the ether, measuring the Earth's velocity in space.
- 1904: Hendrik Lorentz formulates Lorentz transformations. This, as a result of that the Michelson-Morley experiment could not demonstrate the existence of the ether.
- 1905: Albert Einstein published the special theory of relativity.

[0002]

The inventor assumes that the Michelson-Morley experiment and the apparatus used, Michelson interferometer, were not suitable for this. The inventor has made his own calculations that show that regardless of the Earth's velocity in space, regardless of the orientation of the apparatus, the difference between these part experiments was in the order of 30 nanometers! This was within the margin of error and therefore the Michelson-Morley experiment got so-called "zero" result, or "negative" result.

[0003]

The inventor believes that the negative result has been caused by the apparatus design: two arms of about 10 meters each, which were mounted at right angles to each other; three mirrors, one of the semi-transparent. The light beam was reflected in these mirrors or passed these mirrors.

[0004]

The inventor believes that the mathematical and physical analysis of the path of light rays was insufficient and in fact incorrect!

[0005]

Therefore, the device (1ANM) is simple and this means that there are no doubts as to how the light moves inside the device (1ANM). Even the mathematical and physical analysis of the path of the light beam becomes simple, as is the calculation and measurement of the absolute velocity in space becomes easier and clearer!

Description

[0006]

The device (1ANM) is based on the principle that the speed and direction of light are independent of the motions of the source and the observer.

[0007]

List of Figures

Fig. 1 - the device (1ANM), consisting of an arm (A), a laser (L), a screen (S)

Fig. 2 - a section through the apparatus (1ANM); the screen (S) is marked with a coordinate system (x-axis, y-axis); the line connecting the laser (L) to the center of the screen (S) represents the third axis (z-axis) of a three-dimensional coordinate system.

Fig. 3 - two simpler representations of the device (1ANM) that will be used in the various measurements / calculations: a dotted line image and a full line image. On the right side of the image, there is depicted vector $\boldsymbol{v}$ for the absolute speed.

Fig. 4 - shows two possible directions for the absolute speed $\boldsymbol{v}$

Fig. 5 - compilation of the results of eight different measurements / calculations

Fig. 6 - shows how to determine the direction of the

absolute speed vector

[0008]
Fig. M1 – measurement / calculation 1
Fig. M2 – measurement / calculation 2
Fig. M3 – measurement / calculation 3
Fig. M4 – measurement / calculation 4
Fig. M5 – measurement / calculation 5
Fig. M6 – measurement / calculation 6
Fig. M7 – measurement / calculation 7
Fig. M8 – measurement / calculation 8

[0009]
The device (1ANM) consists of the following parts, see Figs. 1 and 2:
- an arm (A) on which the other parts are mounted
- a laser (L) that creates a light dot (P) (with about 1 millimeter in diameter)
- a screen (S) on which the dot (P) ends and where the position of the dot (P) can be read in relation to the center of the screen (S)

[0010]
The arm (A) must be attached to a device (similar to a gyroscope or a robot arm) that allows the arm (A) of the device (1ANM) to be directed, rotated in all possible directions in space.

This part is not part of our invention. There are various technical possibilities to solve this.

[0011]
The distance between the laser heads (L) and the screen (S) must be at least 10 meters.

[0012]
Detailed description of the device (1ANM) components.

[0013]
Arm (A):
At one end of the arm (A) a laser (L) is mounted. At the other end of the arm (A) a screen (S) is mounted. The length of the arm (A) should be such that after mounting the laser (L) and the screen (S), the distance between the laser (L) head and the screen (S) should be at least 10 meters. The arm (A) should be of square cut, and should be designed so firm and stable that no bending of the arm occurs. It is not essential in the context of which material the arm (A) is made.

[0014]
Laser (L):
Laser (L) should be able to transmit a light beam, which is imaged on the screen (S) as a dot (P) of about 1 millimeter in diameter. It is not essential in the context which type of laser is used.

[0015]
Screen (S):
The screen (S) should allow the dot (P) position to be read. The screen (S) should be of square shape, preferably giving the possibility to digitally read the dot (P) position in the coordinate system (x, y) and send this position to a computer for further processing. It is not essential how the screen (S) is manufactured. The size of the screen (S) (readable part) should be at least 6x6 centimeters. This ensures that the light dot (P) motion on the screen (S) does not fall outside. The maximum calculated speed is approx. 850 km/s. This corresponds to the maximum distance of the dot (P) to the center of the screen (S) to about 3 centimeters.

[0016]
Measurements / Calculations:

[0017]
We will use the following terms:
- ***D*** = the length between the laser (L) and the screen (S) (10 meters)
- ***c*** = light speed (300,000 km/s = 300,000,000 meters / second)
- ***v*** = the speed at which the Earth moves in space and with that also the device (1ANM)
(eg 30 km/s = 30,000 meters/second); it is the speed at

which the point on the Earth's surface moves, the point at which the apparatus (1ANM) is located;
it is this speed that will be read / calculated using the device **(1ANM)**

- ***t*** = the time the laser beam needs to reach the screen (S)
- ***d*** = distance between the dot (P) and the center of the screen (S)
- ***x*** = in Fig. M1-M8, ***x*** is a variable used in the calculation

(should not be confused with x-coordinate)

[0018]
We present eight part experiments/measurements / calculations to show how the device (1ANM) moves in space, how the laser beam moves. We make calculations to show where somewhere on the screen (S) the light dot (P) ends up depending on the device's (1ANM) orientation in space and its speed ***v***.

[0019]
These 8 part experiments are shown in Fig. M1-M8. In the upper part we show the original position, depicted with dotted line. Below, there are two overlapping (partially overlapping) images on the device (1ANM), one with dotted lines for the initial

position and one with full line for end position, then when the laser beam reaches the screen (S).

[0020]
To the right of the device (1ANM) is shown the vector for the speed ***v*** (which direction ***v*** has and which orientation the apparatus (1ANM) has against this vector).

[0021]
A point on the Earth's surface has the following motions:
1) motion around the Earth's axis; at equator, the speed is approx. 0.5 km/s
2) Earth's motion around the Sun; the speed is approx. 29.8 km/s
3) The motion of the solar system around the center of the galaxy; the speed is approx. 220 km/s
4) The galaxy moves toward the Great Attractor; the speed is approx. 600 km/s
5) The big attractor, in turn, moves toward the Shapleys superhope, which is a collection of over 8,000 galaxies (not included in my calculations).

[0022]
This means that the absolute speed can be at least approx. 350 km/s
(600-220-30).

With this speed will be ***max(d) = 11 millimeter***.

[0023]
An example of measurement / calculation:
We measure the maximum distance between position of the light dot (P) and the center of the screen (S) ***d***
d = 7 millimeter = 0,007 meter
v = cd/L = 300 000 000 m/s * 0,007 m / 10 m
v = 210 000 m/s = 210 km/s

[0024]
We now present eight measurements / calculations corresponding to eight different relative positions of the apparatus (1ANM) arm (A) with respect to the vector of speed ***v***.
These part experiments / measurements / calculations predict that the apparatus (1ANM) is positioned such that y-axis, z-axis and vector *v* are in the same plane and that when the apparatus (1ANM) rotates 360° in this plane, around the x-axis .

[0025]
Measurement / calculation 1, Fig. M1:
The device (1ANM) moves parallel to the vector ***v****, to the right.*
While the laser beam reaches the screen (S), the entire apparatus (1ANM) moves with distance ***x*** *to the right.*

In this case, the laser beam needs to pass distance D–x.

The light dot (P) ends up exactly in the center of the screen (S).

$$t = x/v = (D-x)/c$$
$$x = Dv/(c+v)$$
$$d = 0$$

[0026]
Measurement / calculation 2, Fig. M2:

The device (1ANM) moves to the right and upwards and forms an angle of 45° with the vector $\boldsymbol{v}$.
The speed is $v_1 = v/2^{1/2}$ *(root of 2).*
While the laser beam reaches the screen (S), the entire apparatus (1ANM) moves by distance x *to the right and by the same distance* x *upwards.*
In this case, the laser beam needs to pass the distance $D-x$.

The dot (P) falls down on the center of the screen (S).

$$t = x/v_1 = (D-x)/c$$
$$x = Dv_1/(c+v_1)$$
$$d = x$$

[0027]
Measurement / calculation 3, Fig. M3:

The device (1ANM) moves upwards and forms an angle of 90 ° with the vector ***v****.*
While the laser beam reaches the screen (S), the entire apparatus (1ANM) moves with distance ***x*** *upwards.*
In this case, the laser beam needs to pass the distance D.

The dot (P) falls down on the center of the screen (S).

$\boldsymbol{t = x/v = D/c}$
$\boldsymbol{x = Dv/c}$
$\boldsymbol{d = x}$

[0028]
Measurement / calculation 4, Fig. M4:

The device (1ANM) moves up and to the left and forms an angle of 45° with vector ***v****.*
The speed is $\boldsymbol{v_1 = v/2^{1/2}}$ *(root of 2).*
While the laser beam reaches the screen (S), the entire apparatus (1ANM) moves by distance ***x*** *upwards and by the same distance* ***x*** *to the left.*
In this case, the laser beam needs to pass the distance ***D+x****.*

The dot (P) falls down on the center of the screen (S).

$$t = x/v_1 = (D+x)/c$$
$$x = Dv_1(c-v_1)$$
$$d = x$$

[0029]
Measurement / calculation 5, Fig. M5:

*The device (1ANM) moves left and parallel to vector **v**. While the laser beam reaches the screen (S), the entire device (1ANM) moves distance **x** to the left. In this case, the laser beam needs to pass the distance **D+x**.*

The light dot (P) ends up exactly in the center of the screen (S).

$$t = x/v = (D+x)/c$$
$$x = Dv(c-v)$$
$$d = 0$$

[0030]
Measurement / calculation 6, Fig. M6:

*The device (1ANM) moves left and down and forms an angle of 45° with vector **v**.*

The speed is $\mathbf{v_1 = v/2^{1/2}}$ *(root of 2).*

While the laser beam reaches the screen (S), the entire apparatus (1ANM) moves with distance $\mathbf{x}$ *to the left and with the same distance* $\mathbf{x}$ *down.*
In this case, the laser beam needs to cut off the distance $\mathbf{D+x}$*.*

The light dot (P) ends up on the center of the screen (S).

$$t = x/v_1 = (D+x)/c$$
$$x = Dv_1(c-v_1)$$
$$d = x$$

[0031]
Measurement / calculation 7, Fig. M7:

The device (1ANM) moves downwards and forms an angle of 90° with the vector $\mathbf{v}$*.*
While the laser beam reaches the screen (S), the entire apparatus (1ANM) moves with distance $\mathbf{x}$ *downwards.*
In this case, the laser beam needs to pass the distance $\mathbf{D}$*.*

The light dot (P) ends up on the center of the screen (S).

$$t = x/v = D/c$$
$$x = Dv/c$$
$$d = x$$

[0032]
Measurement / calculation 8, Fig. M8:

The device (1ANM) moves to the right and down and forms an angle of 45° with the vector $\boldsymbol{v}$.
The speed is $\boldsymbol{v_1 = v/2^{1/2}}$ *(root of 2).*
While the laser beam reaches the screen (S), the entire apparatus (1ANM) moves with distance $\boldsymbol{x}$ *to the right and with the same distance* $\boldsymbol{x}$ *down.*
In this case, the laser beam needs to pass the distance $\boldsymbol{D-x}$.

The light dot (P) ends up on the center of the screen (S).

$$t = x/v_1 = (D-x)/c$$
$$x = Dv_1(c+v_1)$$
$$d = x$$

[0033]
In Fig. 5 we show a summary of the above eight experiments/measurements/calculations and over the position of the light dot (P) on the screen (S) during a

complete rotation of the apparatus (1ANM) by ***360°***.
The calculations show that the distance of the light dot (P) to the center of the screen (S), ***d***,
would vary between 0 millimeters and 1 millimeter (if ***v = 30 km/s***).

[0034]
Fig. 6 shows how the direction of **the absolute speed** is determined.
After finding the maximum distance ***d*** that the dot (P) has towards the center of the screen (S), the vector between position of the dot (P) and the center of the screen (S) determines the direction of **the absolute velocity** in space.

[0035]
In this way, one can determine both the **scalar value** ***v*** for the absolute velocity in space and its **direction**.

2) 1ANM, Claim

C L A I M

Apparatus for measuring the absolute velocity in space -
1ANM (One Arm, No Mirrors)

What I claim as my invention is:

Claim 1.
An apparatus for measuring the absolute velocity in space. The apparatus consists of an arm (A), a laser (L) and a screen (S) arranged so that the laser (L) is mounted at one end of the arm (A) and the screen (S) is mounted at the other end of the arm (A). Laser (L) transmits a laser beam to the screen (S) where it is projected as a dot (P). The position of the dot (P) is read against the center of the screen (S).

The device (1ANM) is mounted on a device that allows the arm (A) to be rotated in all possible directions in space. This device is not included in the patent claim because there may be several technical solutions to achieve this (robot arm; gyroscope).

The device (1ANM) is rotated until the y-axis and z-axis land in the same plane as the vector for speed $\boldsymbol{v}$ (Earth's velocity in space). The method by which this is achieved is not included in the claim because it is dependent on the device on which the arm (A) is mounted.

Claim 2.

This claim consists of the description of the exact path of light through the apparatus, the calculations of the scalar absolute speed, the method of determining the direction of absolute velocity in space. The apparatus (1ANM) together with the description, calculations and measurement method represents the simplest apparatus for measuring of the absolute velocity in space and its direction.

When the apparatus's y-axis, z-axis and the velocity vector $\boldsymbol{v}$ are in the same plane, the measuring method is applied, which is **characterized** in that the arm (A) is rotated 360° in the same plane and that the maximum distance $\boldsymbol{d}$ between the laser beam dot (P) and the center of the screen (S). The greatest distance is obtained when the arm (A), z-axis, constitutes an angle of 90° to the vector $\boldsymbol{v}$ of the velocity. Then the calculation is **characterized** by the following formula

$$v = dc/D$$

where
- v is the absolute speed at which the Earth moves in space (the point on Earth where the apparatus is)
- d is the largest distance between the position of the dot (P) and the center of the screen (S)
- c is the speed of light (in this case in air)
- D is the distance between the laser head (L) and the screen (S)

3) 1ANM, Summary

SUMMARY

Apparatus for measuring the absolute velocity in space -
1ANM (One Arm, No Mirrors)

Scope: Physics.

The device (1ANM) is based on the principle that the speed and direction of light are independent of the motions of the source and the observer.

The device (1ANM) consists of an arm (A), a laser (L) and a screen (S). The device (1ANM) is mounted on a device that allows the arm (A) to be rotated in all possible directions in space. Laser (L) transmits a laser beam to the screen (S) where it is projected as a dot.

The **value of the absolute speed *v*** is calculated using the maximum distance ***d*** between the position of the dot (P) and the center of the screen (S), which is read at a full rotation of the apparatus (1ANM) by 360°.

This value is calculated by the formula ***v* = *dc/D*.**

The **direction of the absolute velocity *v*** in space is determined by the vector created between the light dot (P) and the center of the screen (S) and read on the y-axis at which the reading of the maximum distance of the dot (P) to the center of the screen (S) was made.

4) 1ANM, Drawings

Presented on the following 6 pages.

Note that the figures in this chapter have their own numbering

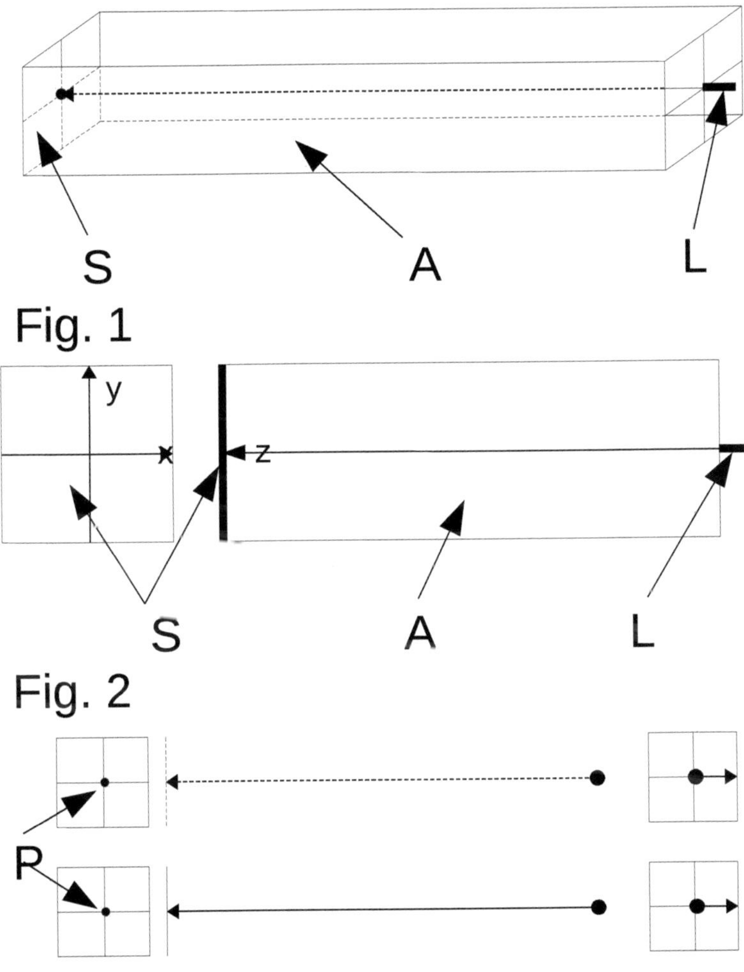

Fig. 1

Fig. 2

Fig. 3

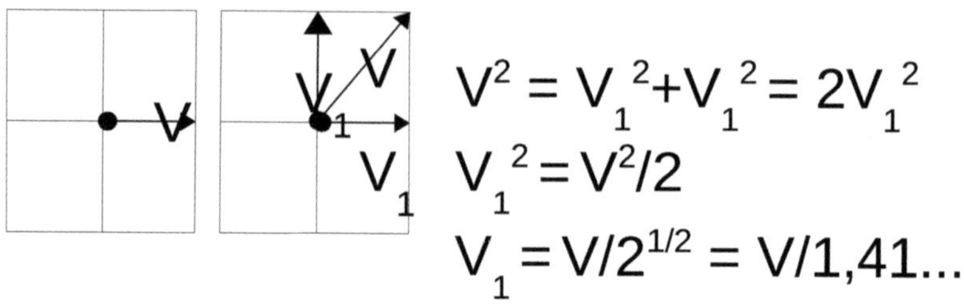

Fig. 4

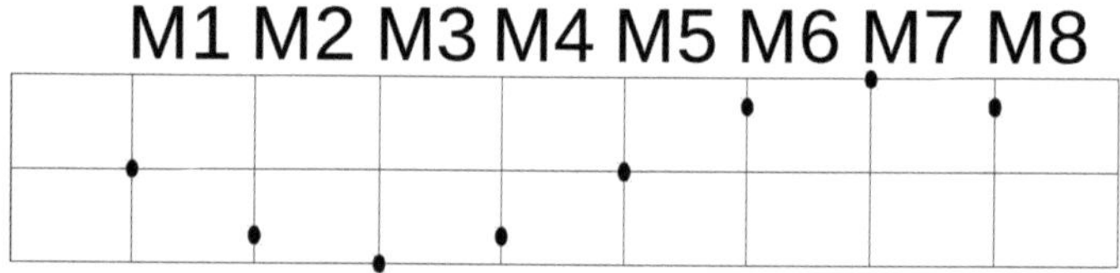

Fig. 5

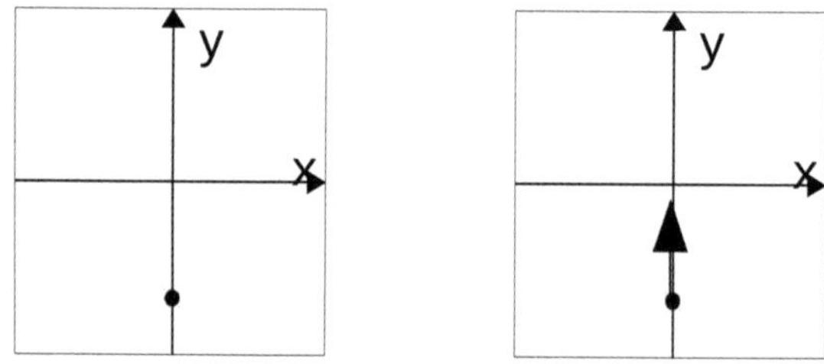

Fig. 6

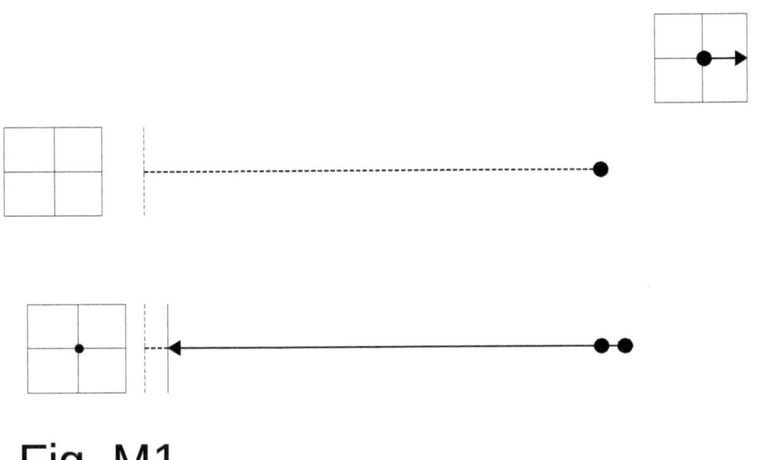

Fig. M1

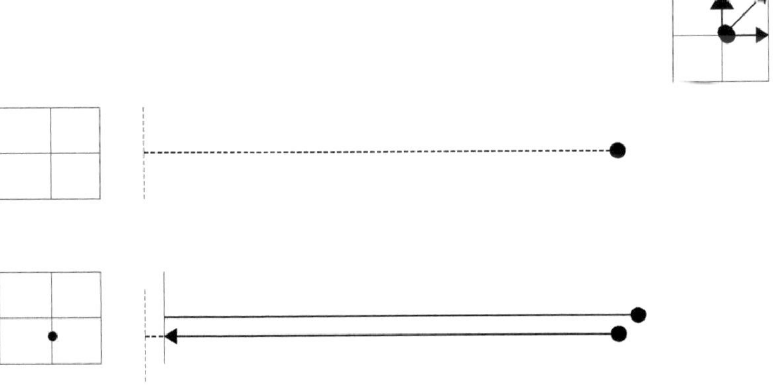

Fig. M2

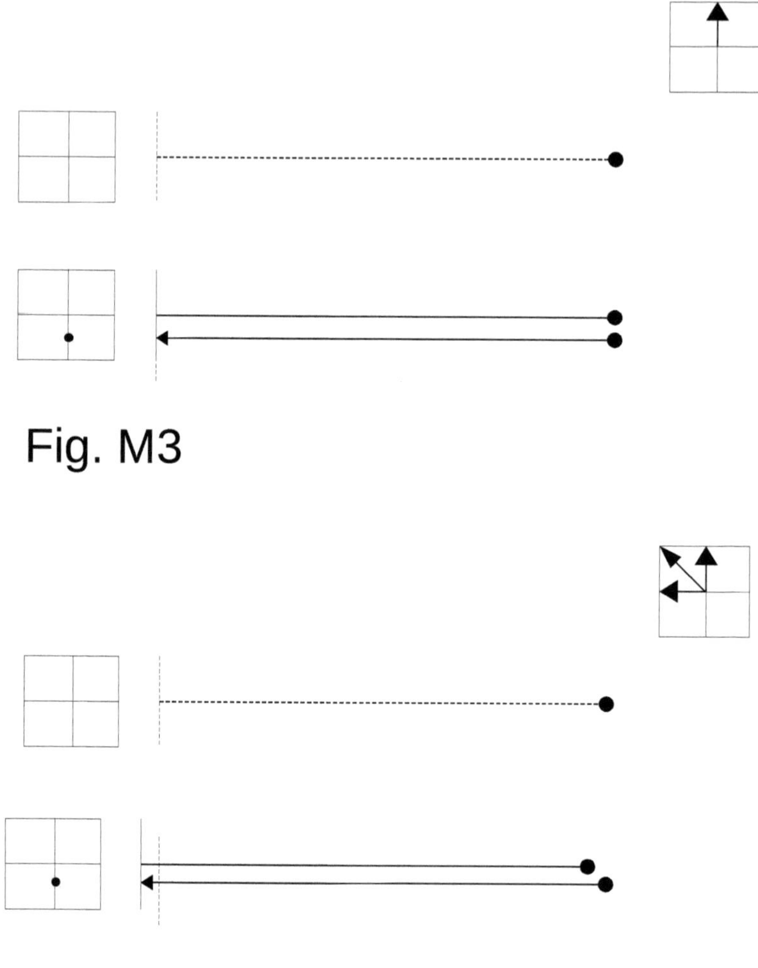

Fig. M3

Fig. M4

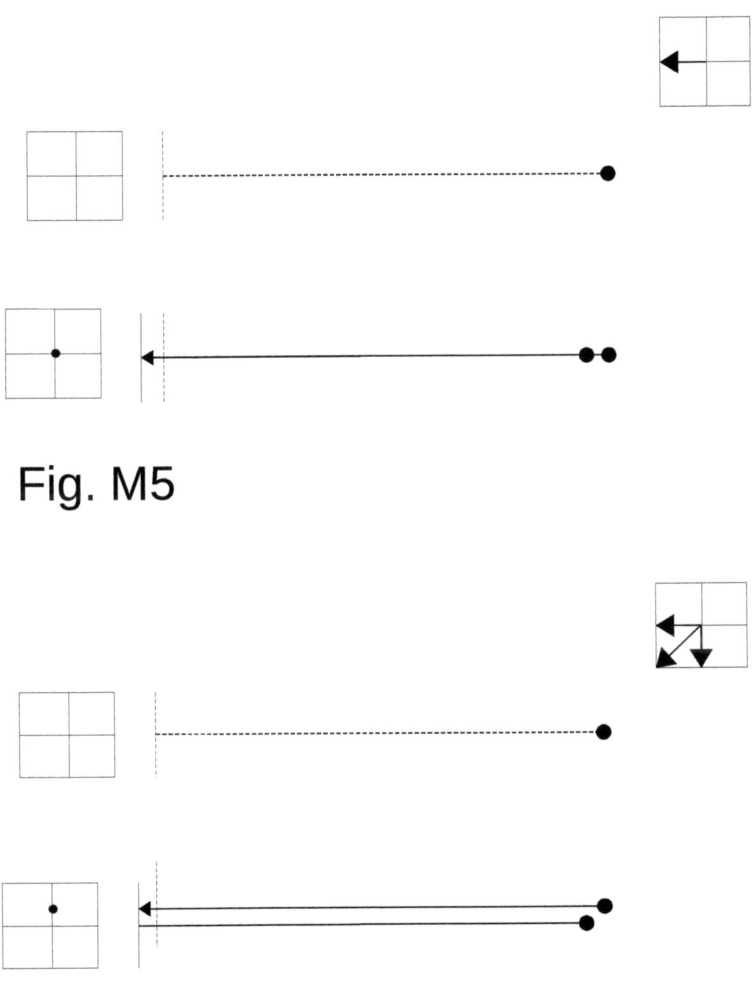

Fig. M5

Fig. M6

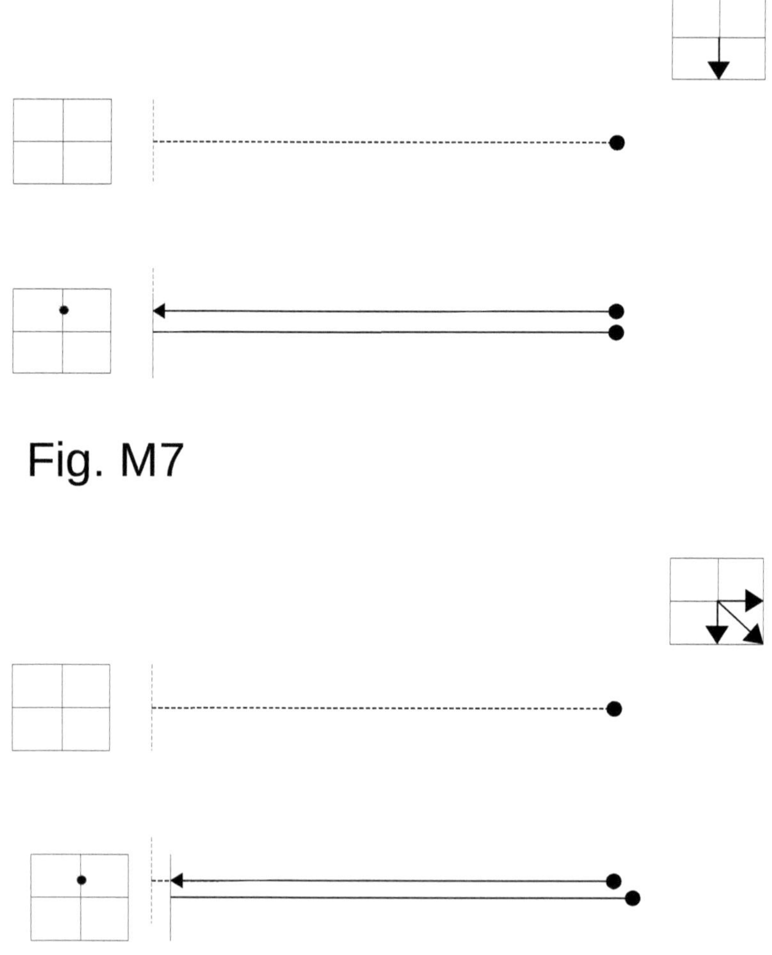

Fig. M7

Fig. M8

Published articles

1)

Physics Essays

Mathematics shows that the Lorentz transformations are not self-consistent

http://physicsessays.org/browse-journal-2/product/1770-3-jan-slowak-mathematics-shows-that-the-lorentz-transformations-are-not-self-consistent.html

2)

SCIREA Journal of Physics

Lorentz Transformations And Time Dilation Do Not Verify Reality

http://www.scirea.org/journal/PaperInformation?PaperID=3699

3)
SCIREA Journal of Physics

Lorentz Transformations - The Sound versus The Light

http://www.scirea.org/journal/PaperInformation?PaperID=3718

14. Epilogue

The special theory of relativity has many unexplained contradictions, inaccuracies, and paradoxes.
It all depends on how you look at the problem.

But if a theory is correct, it should not matter how one views it. It should always be right.
No matter how you turn things around, you should not come to contradictions, inaccuracies or paradoxes!

James Clerk Maxwell determined the speed of light $c = 1 / (\mu_0\varepsilon_0)^{1/2}$. **This is the speed of light in vacuum. This speed is independent of some observers. This speed is independent of the state of motion of the light source.** This is the speed of light in medium in which light propagates. This speed is independent of whether there are any inertial reference systems or not. In the same way as the speed at which the rings on the water propagate independently of any observer on the beach or in any boat. The circles propagate in the water at the same speed. In the same way that the sound propagates in the air regardless of whether one or more people are listening.

In the same way, the speed of light is always equal to

c, regardless of whether there is one or more reference systems that measure this speed, regardless of whether these reference systems are in motion towards each other or if they are at rest towards each other.

Maxwell denominate the medium in which light propagates to the light-bearing ether.

Albert A. Michelson tried to determine the state of this ether, but the 1887 experiment was a complete failure.

Hendrik A. Lorentz tried to explain this failed result and explained this result using two equations that bear his name, so-called Lorentz transformations.

Albert Einstein included these equations in his new theory, **the special theory of relativity**.

Here Einstein, and many others, make so-called derivations of Lorentz transformations. You want to show how all concepts are connected.

We consider two inertial reference systems, S and S', and a light source from which a light signal is emitted. The place where the light source is located and the time when the light signal occurs represent the coordinates of this event.

Both reference systems make their measurements, read their clocks and so on. Each reference system records the place and time of the event with its own name:
(x, t) and (x', t').
The Lorentz transformations then says that between the coordinates of the event in the two reference systems there are the following relations:

$$x' = (x - vt)\gamma \qquad \text{(LTx')}$$
$$t' = (t - vx / c^2)\gamma \qquad \text{(LTt')}$$

where $\gamma = 1 / (1 - v^2 / c^2)^{1/2}$ is called the Lorentz factor.

But this whole theory, SR, basically has a wrong view of what light is and how the two reference systems and light relate to each other!

It is said that the speed of light should be the same in all reference systems, that the light moves so and so in each reference system.

But the reality is different from what is considered within SR. The light has the same speed in vacuum, in the light-carrying ether, everywhere and always!
And it is actually the two reference systems that move in the vacuum as well, in the light-carrying ether!

Both the two reference systems and the light signal move in a vacuum reference system.

An observer does not know his own absolute velocity in space and its direction. This speed has not been used in any of the experiments that have been done so far, experiments where one examines the speed of light or other phenomena where the speed of light is included as a component.

Therefore, when an observer in an inertial reference system analyzes, measures, the speed of light, it is about **the relative speed** between the reference system and the wavefront of the light signal!

That is why a concrete experiment will never bring out the absolute speed of light!
That's why it was only Maxwell who got the right value in an empirical way!

We illustrate this in the following image. There we depict three inertial reference systems, one that is in absolute rest in vacuum (can you name something like that?), the second that has the absolute velocity in vacuum equal to v', and the third that has its absolute velocity in vacuum equal to v'', $v'' > v'$.

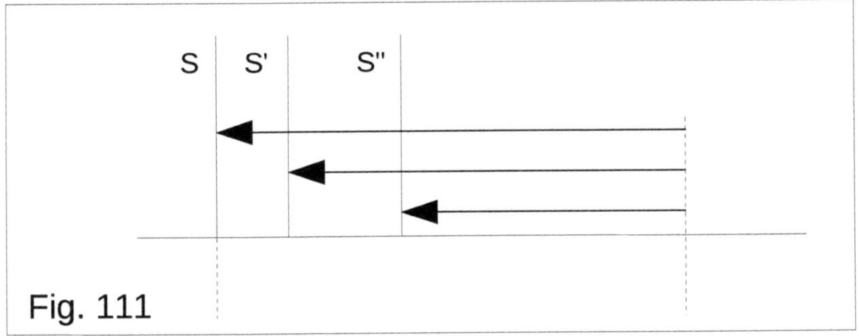

Fig. 111

In the beginning, the three reference systems are at the same point. Then a short light signal occurs on the x-axis.
The light signal will propagate at the same speed *c*, regardless if there are some reference systems there or not, regardless of whether they move or not, regardless off at what speed they move.

The light signal will reach first S'', then S' and finally S. This means that the **relative speed** between the three reference systems and the light signal is:
For S: $\boldsymbol{c}$, for S': $\boldsymbol{c + v'}$, for S'': $\boldsymbol{c + v''}$.

But is the picture Fig. 111 correct? No! Not really!

We look at the picture Fig. 112.
This image depicts three different moments of reality.

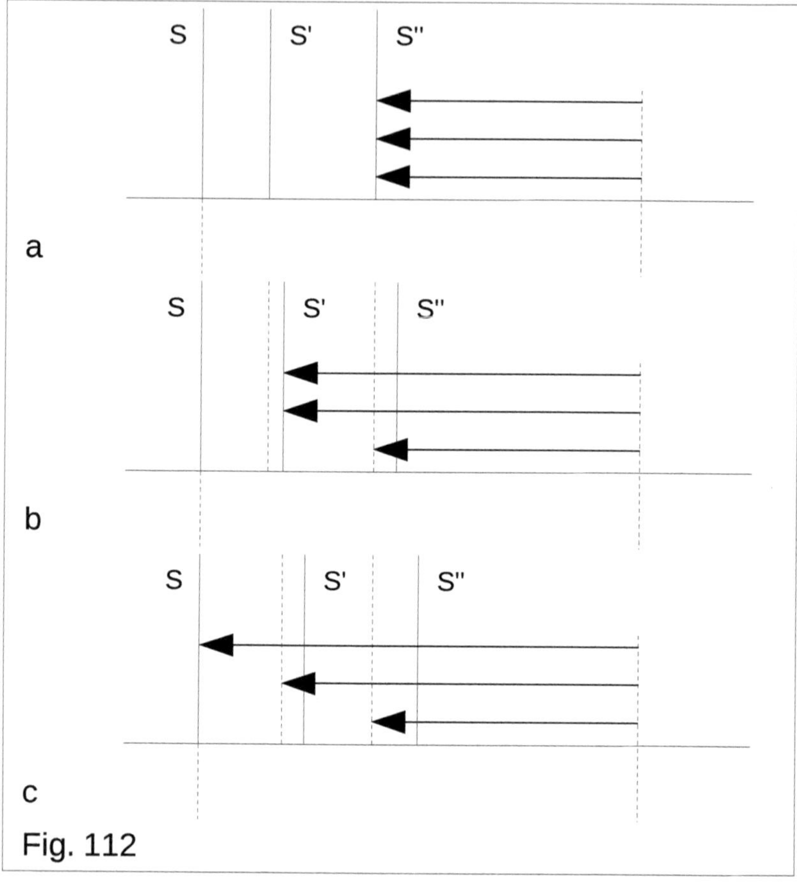

Fig. 112

Fig. 112-a depicts the moment when the light signal reaches S''.
Fig. 112-b depicts the moment when the light signal reaches S'.
Fig. 112-c depicts the moment when the light signal reaches S.

The time t'', the moment when the light signal reaches S'', helps us to calculate the distance between S-origo and S''-origo, and the distance between S''- origo and E (event), at exactly that moment!

The time t', the moment when the light signal reaches S', helps us to calculate the distance between S-origo and S'-origo, and the distance between S'- origo and E (event), at exactly that moment!

The time t, the moment when the light signal reaches S, helps us to calculate the distance between S-origo and E (event), at exactly that moment!

See calculations in
SCIREA Journal of Physics
Lorentz Transformations And Time Dilation Do Not Verify Reality

Think about how it is in the three chapters with water, air, and vacuum.

Therefore, the statement from SR, that *the light speed is the same in all reference systems*, is incorrect!

Therefore, one gets contradictions when analyzing the derivation of Lorentz transformations.
Therefore, the whole special theory of relativity is nonsense!

How do researchers react to this? Why do universities and physical institutions ignore this? Why do you teach SR? Why do you teach a wrong theory?

Jan Slowak: Light - The Absolute Reference in the Universe